TRANSPORTATION ENGINEERING BASICS

Second Edition

A.S. NARASIMHA MURTHY
HENRY R. MOHLE

ASCE PRESS

American Society of Civil Engineers
1801 Alexander Bell Drive
Reston, Virginia 20191–4400

Abstract: This revised edition of *Transportation Engineering Basics* presents 22 labs focusing on traffic control devices, travel time and delay studies, saturation flow rates, sight distance and gap study, level of service analysis, and parking studies, as well as transportation planning and traffic engineering, transportation modeling, transportation funding, and intelligent transportation systems (ITS). This edition also includes the latest information about computer software applications, Internet use in the field, and a list of practical print sources.

Library of Congress Cataloging-in-Publication Data

Murthy, A. S. Narasimha
    Transportation engineering basics / by A.S. Narasimha Murthy and R. Henry Mohle.--
Rev. 2nd. Ed.
        p. cm.
    Includes bibliographical references and index.
    ISBN 0–7844–0464-X
    1. Transportation engineering. I. Mohle, R. Henry. II. Title.

TA1145 .M88 2000
388—dc21

00-063965

Dedicated to my
late father A.L. Sitaram,
my mother A.S. Lakshamma,
and my family

A.S. Narasimha Murthy

PREFACE TO THE FIRST EDITION

The main purpose in writing this manual was to develop a tool that clearly illustrates the basics of the transportation and traffic engineering field for undergraduates and new recruits. Another objective was to reinforce academic concepts and principles by providing practical exposure to the transportation field. This manual is a quick guide for those who have some advanced knowledge and experience in the area. It also attempts to answer the questions most commonly asked by undergraduate students about what transportation and traffic engineers solve in everyday work. The manual has been designed for use as a text both for workshops and for undergraduate lab classes.

Since transportation engineering is a practical area of study, most of its aspects are difficult to simulate in a classroom. Everything that happens in this field takes place outside the classroom. On-site experience is the key element in understanding and finding suitable solutions to transportation problems. Transportation engineering incorporates several fields of engineering and science. This manual exposes students and new recruits to the various areas of application required to solve problems in transportation engineering. Examples range from the application of queuing theory and determining the length of a left-turn lane at an intersection to understanding the socioeconomic and traffic impacts of a proposed shopping mall on surrounding areas.

This manual attempts to answer another question typically asked by undergraduate students considering transportation engineering as a major study area – "What type of work will I be doing?" The types of work that newly recruited transportation engineers are expected to perform are illustrated here. The manual introduces the technologies of the 1990's with emphasis on latest codes and standards for easy reference. Students are made aware of what they are expected to do at a workplace.

The labs are structured so those students in a workshop or lab can complete assignments the same day, with only one or two additional hours for calculations and preparing the reports. The worksheets were developed specially for on-site recording of field data based on the authors' practical experience. This manual was also prepared considering the needs and constraints of teachers. In our combined experiences in teaching at various levels, the first two classes of any workshop or lab are the most difficult ones for teachers to organize. In view of this, the first two chapters of this manual require less time and preparation to begin the classes.

The material in this manual has been prepared so that it is self-explanatory, and all the lab work can be conducted without any need for major equipment. To ensure that each student participates in the fieldwork, an examination is suggested after the lab sessions. In conducting the labs it is recommended that students be grouped and given different sites to collect data from, for comparison purposes.

In conducting a workshop or lab it is suggested that the teacher in charge take the class on a tour of the local transportation agency. This will help familiarize students with the types of work (studies and analyses) conducted at the workplace, and it will also help demonstrate the job opportunities available in the transportation field. There are many local transportation agencies that willingly cooperate, conducting tours of their operations for students.

We are extremely grateful to the American Society of Civil Engineers and ASCE Press for having the confidence in our ability to prepare this important manual. Our special thanks go to Ms. Zoe G. Foundotos, who has been of great help from the very beginning in preparing this manual. Thanks also are extended to Dr. Kumares Sinha, Mr. T.C. Sutaria, Dr. Edward S. Neumann, and Mr. Kenneth O. Voorhies for their input throughout the review process.

Our special thanks to Dr. Donna Nelson for suggesting several important changes to the format and text writing. Also, we thank Mr. Chalapathi Rao Sadam for editing the manuscripts. We sincerely thank Mr. Ali Nazem for preparing outstanding figures, tables, and worksheets on Auto CAD. Our thanks to Ms. Helen Schefcick, who spent numerous hours during weekends and holidays typing the manuscripts. This manual could not have been prepared without the help of Mohle, Grover & Associates, which provided computer and other needed facilities to complete this important work.

A.S. Narasimha Murthy and Henry R. Mohle

PREFACE TO THE SECOND EDITON

Ten new labs are added to this revised edition, and should provide both undergraduate and graduate students with advanced practical knowledge in planning and operational elements of transportation. This edition covers multimodal (auto, transit, bicycle, and pedestrian) aspects of transportation. The new areas are modeling, air quality, transportation demand management, and transportation system management. These new labs are added to show students how planning and operations are used in combination to resolve transportation system problems. New labs on transportation performance monitoring and transportation politics are included to show the depth and breadth of the transportation field, not just projects.

The objectives of the labs are the same as in the first edition: to provide a preliminary practical experience and to show the students how classroom knowledge is applied in working places. The labs are designed to allow the students to think about how to obtain data and solve transportation problems. For example, the "trip diary" assists in understanding the transportation planning process and its use in travel demand modeling. The lab on air quality provides basic knowledge about air quality, and the importance of protecting in environment. The lab on transportation funding and politics provides insight into whom makes the decision regarding transportation projects.

The questions at the end of each session assist students in contacting public agencies to obtain necessary data. The questions at the end of each lab test the computer application abilities and allow learning the communication (oral) capability of students. In addition, by contacting the agencies, students get an insight into different types of jobs available in the transportation field.

In the six years since the first edition of *Transportation Engineering Basics* was published, several major changes to both planning and operations have taken place. The changes are mostly attributed to the passing of Intermodal Surface Transportation Efficiency Act (ISTEA, 1990) and Transportation Equity Act for the 21st Century (TEA-21, 1998). Such legislation has provided opportunities for transportation professionals in terms of new funds and jobs. At the same time, new thinking and new concepts have been also developed. For example, the passing of ISTEA required that transportation planning projects show the benefits of a project including monitoring performance as a part of the project.

The web sites included in the appendices provide information about transportation sites that offer vast information on transportation planning, operations, and job sites; some sites provide links that are useful in further searches. The section on software use in transportation has been revised to reflect current availability, as well Internet access. We thank Lawley Publications for allowing us to reprint some of their researched tables from *The Urban Transportation Monitor*.

We once again thank the American Society of Civil Engineers (ASCE) and ASCE Press for giving us an opportunity and showing confidence in our work. We sincerely thank Joy Bramble of ASCE Press for her patience and guidance throughout the production of this revised edition. We thank our behind-the-scenes reviewers, who provided essential input to improve this edition. We thank Ali Nizem and Joe Pontejos for helping us with figures and Arvind Murthy for providing assistance with computer software and figures. We thank Mr. Salvador Cortez for developing the cover for this book.

A. S. Narasimha Murthy and Henry R. Mohle

TABLE OF CONTENTS

LIST OF FIGURES

LIST OF TABLES

WORKSHEETS

Lab 1

Traffic Control Devices Inventory

OBJECTIVES

Objectives of this lab are to identify, classify, and record various traffic control devices (signs) installed along survey route(s) during off-peak hours.

NEED FOR THE STUDY

Traffic control devices are installed along travel routes to ensure safe, orderly, and predictable traffic movement. They provide guidance and warnings to the motorists.

OVERVIEW

The *Manual on Uniform Traffic Control Devices* (*MUTCD*) was developed to promote uniformity in traffic control devices. *MUTCD* presents traffic control device standards for all streets and highways open to public travel, regardless of type or class, or of the governmental agency having jurisdiction.

According to *MUTCD*, installation of a traffic control device must satisfy five requirements; a device must:
Fulfill a need
Command attention
Convey a clear, simple meaning
Command respect of road users
Give adequate time for proper response

Installation of a traffic control device is subject to design, placement, operation, maintenance, and uniformity. Design ensures that size, contrast, color, shape, and composition are acceptable and serve the intended purpose. Placement is particularly important, because vehicle drivers, traveling at normal speed, need adequate time to read and comprehend the sign, and to make any necessary changes required to heed the sign. Operation, maintenance, and uniformity are also important, because drivers must be able to recognize the sign in order to react to it. Figure 1-1 shows sample traffic control devices.

1

STUDY COMPONENTS

In conducting a traffic control device inventory, signs are identified, classified, and recorded, including: regulatory signs, warning signs, guide signs, .recreational and cultural signs, and construction and maintenance signs.

Regulatory Signs

Regulatory signs include right-of-way (stop and yield), speed, movement (directional), parking, pedestrian, and other special signs. A specific code is assigned to each type of sign on the street plans for easy recognition. For example, a stop sign is coded R-1. On a road plan, all R-1s mean stop signs. Regulatory signs are usually rectangular with a black legend on a white background.

Warning Signs

Warning signs alert drivers to any potentially hazardous condition on or adjacent to the roadway provide a signal to reduce speed or drive safely. Warning signs are used to indicate intersections, traffic signals, changes in grade, entrances, and crossings. They are usually diamond shaped with black legend and border on a yellow background.

Guide Signs

Guide signs indicate routes and direct travelers to cities, places of interest, parks, forests, and historical sites. Most guide signs vary in size and usually feature white messages on a green background.

Recreational and Cultural Signs

Signs for general motorist services are white messages on a blue background. Recreational and cultural interest area signs, which are included under guide signs, have a white symbol on a brown background.

Construction and Maintenance Signs

Construction and maintenance operation signs are usually black text on an orange background. Special construction and maintenance signs follow the basic standards of shaped and other details for all highways.

FIELD WORK AND DATA COLLECTION

In the field, traffic control device data collection occurs along a designated roadway. Signs installed on both sides of the roadway are recorded. The data collected include type of traffic control device, as well as sign color, shape, code, and condition. For

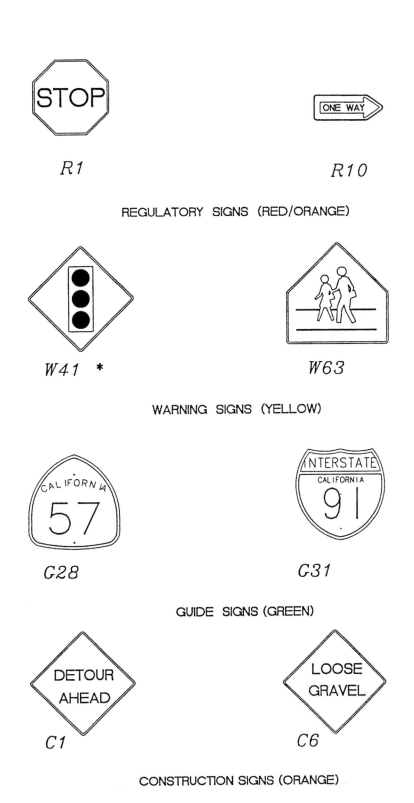

Figure 1-1 Traffic control devices (signs).

future reference, the direction of travel (eastbound, westbound, etc.) is also recorded. In practice, the data collection is done by video logging the street signs. The collected video logged data is analyzed in the office and entered into software developed specially for traffic control devices. This field data is used for maintenance and identification purposes.

FORMULATING CONCLUSIONS AND RECOMMENDATIONS

1. List the type(s) of traffic control devices, including color, shape, and message(s) that are installed along the survey route.
2. List the code(s) used for each sign identified (survey) per *MUTCD*.
3. List the total number of regulatory, warning, guide, and other signs installed along the survey route.
4. Describe the condition of signs installed along the survey route.
5. Describe whether or not the surveyed traffic control devices are consistent with *MUTCD* standards.
6. Describe what improvements (add or remove) are required to the traffic control devices along the survey route (add, remove, or replace).

WORK SHEET: LAB 1

TRAFFIC CONTROL DEVICES INVENTORY (TCDI)

DATE _____ DAY_____ WEATHER _____ PAGE _____

STUDENT NAME(S)/GROUP NUMBER _____

ORIGIN _____ DESTINATION _____

STARTING TIME _____ ENDING TIME _____

SIGN DESCRIPTION	TYPE	CONDITION	PARTICULARS
EXAMPLE STOP SIGN	REGULATORY	GOOD	ON VALLEY BLVD —EASTBOUND (E/B)
ONE WAY	REGULATORY	GOOD	ON VALLEY BLVD AT TEMPLE AV. (E/B)

Lab 2

Travel Times and Delay Study

LAB OBJECTIVES

The main objectives of this lab are to determine the fixed delay, variable delay and total travel time delay on three different routes with the same origin and destination during peak and off-peak hours.

NEED FOR THE STUDY

Travel time is the most important element considered by drivers when taking alternate routes to reach a destination. The pros and cons are checked, usually based on previous experience or based on recent information received either in newspaper or radio about the route(s) conditions. The main purposes of conducting a travel time delay study along a route or network is to identify problem areas, congestion, bottle necks, effectiveness of installed traffic control devices, link travel times, and other travel related characteristics.

GENERAL OVERVIEW

Irrespective of the route selection, "delays" are part of travel and will occur due to certain traffic control devices (fixed delays) and other incidents such as accidents, and construction along the route (variable delays). The summation of fixed delays and variable delays give the total delay between an origin and a destination. The total delay varies depending on the time of travel during peak and off-peak periods. It has been observed that the total delay will be large during peak periods as the volume of traffic is larger than during off-peak periods. Therefore, the selection of a route depends on the time of travel, types of traffic control devices, and previous experience and knowledge of a particular route(s). Figure 2-1 shows three possible routes from an origin to a destination with different road facilities such as freeways, and arterial streets.

STUDY COMPONENTS

The following are the important components to be studied and recorded in conducting a travel time delays study on a specific route:

Fixed delay

It is caused due to traffic control devices and other devices that are in stalled on the route and cause delays irrespective of traffic volume or time of travel.

Variable delay

It is caused along the route during travel due to interference of other traffic on the road. Stopped-time delay is caused when the vehicle is actually standing without moving due to various reasons such as congestion, construction, and accident.

Travel-time delay

This is the difference between the total travel time and the calculated time that is based on traversing the study route at an average speed corresponding to uncongested traffic flow (off-peak hours) on the route.

Delay

It is the total time lost during travel due to fixed and variable delays.

FIELD WORK AND DATA COLLECTION

The delay data is collected on selected alternate routes having same origin and destination. Fixed delay data is recorded on the field whenever the vehicle is stopped due to existing traffic control devices such as traffic signals, stop signs and other traffic control devices that requires the vehicle must make a stop before moving further. The time at the beginning of stop and ending of stop are recorded. In case of fixed delay all vehicles on the road are subjected to this delay.

Variable delay data is recorded on the field whenever delay is caused due to other traffic in the lane or an accident delay or other traffic related reasons not related to fixed delay. It is important to record the actual time spent when the vehicle is at rest (idling). The average speed of travel on the study route is recorded while conducting the delay study.

The total travel time delay to traverse the complete route is determined by summing up both the fixed and variable delay values recorded for that particular route. While conducting a delay study, record the number of traffic control devices, stop signs, and posted speed. Also, record the characteristics of the road such as number of lanes, number of signals, turning lanes at intersections and any construction activities on the survey routes.

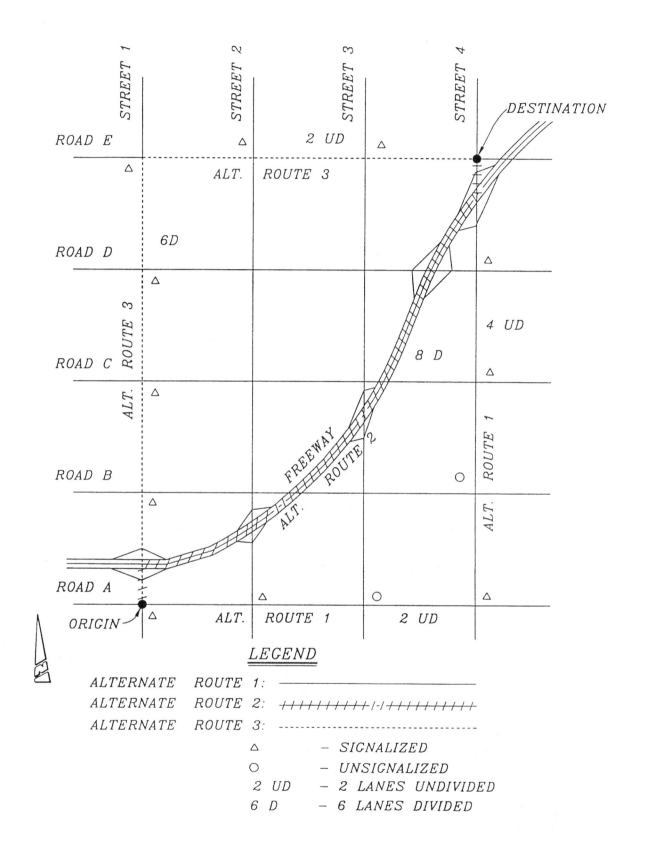

Figure 2-1 Travel time and delay study.

8

FORMULATING CONCLUSIONS AND RECOMMENDATIONS

1) What are the characteristics of survey route?
2) What was the average travel speed on each route surveyed?
3) What was the total fixed delay on each route (sec)?
4) What was the total variable delay on each route and list the reasons?
5) What was the net delay on each route?
6) Which assigned route has the least total delay and why?
7) What improvements are required to reduce total delay on each route surveyed?

WORK SHEET: LAB 2

TRAVEL-TIME AND DELAY STUDY

DATE _____ DAY _____ WEATHER _____ PAGE _____

STUDENT NAME(S)/GROUP NUMBER _____

ORIGIN _____ DESTINATION _____

STARTING TIME _____ ENDING TIME _____

SELECTED ROUTE _____

PEAK PERIODS _____

STARTING POINT	ENDING POINT	DELAY STARTING TIME	DELAY ENDING TIME	TOTAL DELAY (SEC)	CAUSE OF DELAY	TYPE OF DELAY
EXAMPLE * ROAD A (E/B)	STREET 2	3:00:00	3:02:00	120	S	F
		TOTAL DELAY				

SYMBOLS USED S = SIGNAL (TRAFFIC LIGHT), ST = STOP, PK = PARKED CARS, F = FIXED DELAY, PED = PEDESTRIAN, ACC = ACCIDENT, V = VARIABLE DELAY, E/B = EASTBOUND

Lab 3

Spot Speed Studies

LAB OBJECTIVES

The main objectives of this lab are to conduct a spot speed study and develop cumulative speed distribution curve. The speed survey is to be conducted along a straight section of a road during off-peak hours.

NEED FOR THE STUDY

In recording travel time characteristics along a route, one of the key elements recorded is the "speed". Speed of travel on the road is used in classifying routes as freeways, highways, arterials, collectors, and local streets. Level of service based on speed is an indicator of quality of traffic flow or mobility.

GENERAL OVERVIEW

Speed is the ratio of total distance traveled divided by the total time taken to traverse that distance. Units for speed are miles per hour (mph) or feet per second (fps) or meters per second (mps). Speeds vary with the type of road and traffic volume. It is higher along freeways and highways and lower along collector and local streets. Speed is affected by factors such as lane width and sight distance. Speed decreases with an increase in traffic volumes. Traffic volume is defined as the number of vehicles that pass a point along a roadway or traffic lane per unit time.

STUDY COMPONENTS

There are three basic measures of speed. They are 1. Spot Speed; 2. Overall Speed; and 3. Running Speed.

Spot Speed

Spot speed is the speed of the vehicle as it passes a fixed point along a section of the roadway. Spot speed is determined by measuring the time required for a vehicle to traverse a specified distance along a road. Spot speed studies are conducted to draw the speed distribution curve along a road section. Figure 3-1 shows the cumulative spot speed distribution on a roadway.

Overall Speed

Overall speed for a route is obtained by dividing the total distance by the total time (sum of fixed and variable delays, lab 2) of travel.

Running Speed

Running speed is determined by dividing the total distance by the total running time (only the time vehicle is in motion as defined in lab 2) for the route. That is, all stop-time delays are excluded. Overall Speed and Running Speed studies are conducted over a specified route for determining quality of service between alternate routes. The main purpose of all speed studies is to obtain speed distributions, identify hazardous areas (excess speed), accident analyses, traffic control planning, and check geometric design. For speed studies, off-peak hours are used for conducting surveys on open stretches of straight roads away from the influence of stop signs, construction and signals.

FIELD WORK AND DATA COLLECTION

There are several methods by which spot speeds can be recorded. The most commonly used methods are the use of machine recorders, photographic method, and speed radar guns. Also, the spot speeds can be recorded manually. The manual method is described in this lab for use. Figure 3-1 shows the cumulative spot speeds distribution graph.

Spot speed of vehicles passing a straight section of the selected roadway is calculated by recording the time required to cross a specified distance called the "trap" length. For this purpose a straight section of the road is selected for marking the trap length. The section is called as the test section. The recommended trap lengths for manual speed studies are as follows:

Table 3-1 Trap length for spot speed studies.

Average Speed of Traffic (mph)	Recommended Trap Length (feet)
Below 25	88 - 100
25 - 40	176 - 200
Over 40	264 - 300

In collecting the data, an observer starts and stops the stopwatch as the vehicle enters and departs the starting and ending points of the marked test section. The test section is marked as shown in Figure 3-2. The trap length is determined based on the posted speed of the roadway. Transverse tapes are placed on the roadway as shown in Figure 3-2 for marking the test section. In actual study to avoid parallax error endoscopes or flash boxes are used at the end of each trap length. The observer should be in such a position that the on coming drivers are not influenced to change their speed as they approach the test section.

12

A sample size of at least 100 to 150 cars is required to get a good distribution and compensate errors in field data collection. In a platoon of cars approaching the starting point of the test section, take the first car for data and later switch to other cars in the platoon. This method of collecting spot speed data is called as time versus measured distance method. Use work sheets provided for this lab 3 to collect the spot speed data and draw the cumulative distribution graph shown in Figure 3-1, using the survey data.

SPEED DISTRIBUTION CURVE

In Figure 3-1 the vertical axis is the percentage of vehicles traveling at or below the posted speed. In practice the 85th percentile speed is used in recommending the posted speed to either increase or decrease, based on several factors. The 50th percentile speed is an alternative measure of the average speed of travel on the road way. Thus, the spot speed data, when plotted on a graph, indicates the speed characteristics of the roadway section surveyed.

FORMULATING CONCLUSIONS AND RECOMMENDATIONS

1) What are the road characteristics of test section and draw sketch of test section?
2) What was the posted speed and actual trap length used?
3) Draw the speed distribution curve and identify the 50th and 85th percentile speeds?
4) What conclusions can you make from the cumulative distribution curve?
5) Call your local City Traffic Engineering/Public Works Department and request for Spot speed studies report and analyze the data

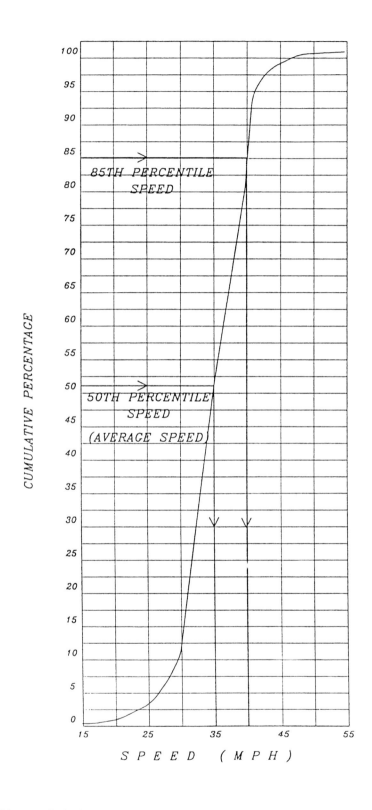

Figure 3-1 Spot speed study cumulative distribution.

14

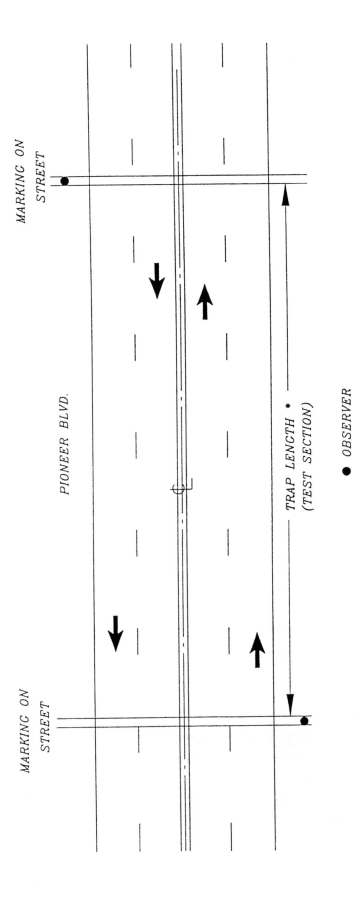

Figure 3-2 Spot speed study: trap length.

15

WORK SHEET: LAB 3

SPOT SPEED STUDY

DATE _____ DAY _____ WEATHER _____ PAGE _____

STUDENT NAME(S)/GROUP NUMBER _____

STREET NAME _____

POSTED SPEED _____ MILES PER HOUR

LENGTH OF SPEED SECTION SELECTED _____ FEET

PEAK PERIODS _____

TOTAL NUMBER OF VEHICLES OBSERVED _____

SPEED GROUPS (COL. 1)	FREQUENCY OBSERVED (COL. 2)	CUMULATIVE FREQUENCY (COL. 3)	COL. 1 × COL. 2 (COL. 4)	COL. 2 × COL. 4 (COL. 5) *
TOTAL				

* Data used for plotting cumulative distribution graph

Lab 4

Turning Movements Count and Peak Hour Factor

LAB OBJECTIVES

The main objectives of this lab are to record turning movement counts at an intersection and calculate Peak Hour Factor. The turning movement counts are to be collected during peak periods and off-peak periods.

NEED FOR THE STUDY

Existing traffic data at an intersection or on a road section forms the foundation for analysis in the transportation field. The existing traffic count data is important for both planning and operational studies. Turning Movement Counts (TMC's) are the directional volume of traffic passing through an intersection. The Average Daily Traffic (ADT) or the 24-hour volume on a road is recorded to study the hourly traffic volume variation on the roadway. Also, to determine the peak period for the road and nearby intersections.

TMC's are usually taken during peak periods at an intersection to conduct Level of Service (LOS) analysis and Peak Hour Factor (PHF). Sometimes traffic flow during off-peak period is also recorded. The recorded peak and off-peak period traffic are compared with the ADT.

GENERAL OVERVIEW

The volume of traffic passing through an intersection is usually recorded as vehicles per hour (vph). TMC's at an intersection relate to the left, right, and through traffic leaving the intersection from each approach. The representation of intersection geometric is as shown in Figure 4-1. Typical directional distribution of traffic flow at a four-way intersection is shown in Figure 4-2.

Turning Movement Counts and Average Daily Traffic

Table 4-1 shows the turning movements taken at a study intersection during p. m peak periods. In practice TMC's are collected for two hours duration at a study intersection to conduct intersection level of service (LOS) analysis. The interval used for collecting data is 15 minutes. The peak hour traffic is defined as the four successive fifteen-minute periods traffic record. The representation of TMC's at study intersections is shown in Figure 4-3.

Table 4-1. Intersection turning movement counts.

N/S Direction: La Brea Avenue; Weather:Good R=Right ; T = Through: L= Left
E/W Direction: 6th Street

Date: August 2, 2000
Day: Tuesday

Start Time	Southbound			Westbound			Northbound			Eastbound			Total
	R	T	L	R	T	L	R	T	L	R	T	L	
a.m.													
7:00	14	232	5	5	98	12	3	202	12	25	56	7	671
7:15	17	274	10	4	110	12	14	212	10	23	65	9	760
7:30	21	331	13	6	155	8	11	225	7	18	62	9	866
7:45	13	402	9	16	230	12	15	269	11	16	51	11	1055
Hour Total	65	1239	37	31	593	44	43	908	40	82	234	36	3352
8:00	21	348	8	8	147	18	5	303	18	38	84	11	1007
8:15	26	411	15	6	165	18	21	318	15	35	98	14	1140
8:30	32	497	20	9	233	12	17	338	11	27	93	14	1299
8:45	20	603	14	24	345	18	23	404	17	24	77	17	1583
Hour Total	98	1859	56	47	890	66	65	1362	60	123	351	54	5028
p.m.													
4:00	13	206	8	3	83	9	11	159	8	17	49	7	570
4:15	16	248	10	5	116	6	8	169	5	14	47	7	650
4:30	10	302	7	12	173	9	11	202	8	12	38	8	791
4:45	21	663	15	26	380	20	25	444	18	26	84	18	1741
Hour Total	60	1419	39	46	751	44	55	973	39	69	218	40	3752
5:00	19	308	11	5	124	14	16	239	11	26	73	10	855
5:15	24	372	15	7	174	9	12	253	8	20	70	10	974
5:30	15	452	10	18	259	14	17	303	12	18	57	12	1187
5:45	32	995	22	40	569	30	37	666	27	40	126	27	2611
Hour Total	90	2128	58	69	1126	66	82	1460	59	104	326	60	5627

18

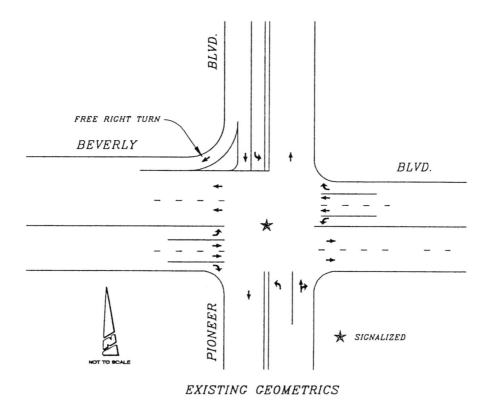

Figure 4-1 Representation of intersection geometrics

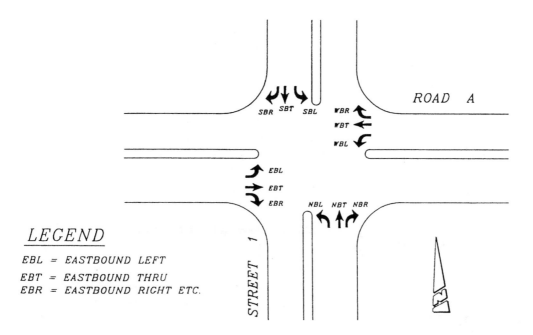

LEGEND

EBL = EASTBOUND LEFT
EBT = EASTBOUND THRU
EBR = EASTBOUND RIGHT ETC.

Figure 4-2 Representation of directional movements.

19

Traffic counts at an intersection are recorded using Manual Count Boards or Electronic Boards. The main advantage of electronic boards is that they can be preset to record the volume of traffic at any required interval. The TMC's are also used for capacity analysis, signal timing, and signal warrants analysis.

Table 4-2 shows the ADT volume collected on a road section for the westbound direction. This data will give the present demand on the street and help in establishing the peak periods at nearby intersections.

Figure 4-4 shows the graphical representation of hourly traffic volume variation on a study road section. The peak periods of the road are clearly observed in Figure 4-4. If large percentage of heavy vehicles such as trucks, buses and recreational vehicles, it is necessary to record them or take vehicle classification counts.

Peak Flow Rate and Peak Hour Factor

The traffic volume is defined as the total number of vehicles passing a point on a lane or roadway during a specified interval of time. The volume of traffic is expressed as ADT, Annual Average Daily Traffic (AADT), and vph. The rate of flow of traffic is defined as the equivalent hourly rate at which vehicles pass over a given point or section of a lane or roadway during a given time interval less than one hour. That is, if counts are taken for less than one hour, using suitable factors, it is expanded to represent one-hour volume. For example, if 15 minutes count is conducted at an intersection, the volume recorded is multiplied by four to obtain the hourly volume.

The peak flow rate plays an important role in the design of traffic control devices. When turning movement counts are taken at an intersection, the counts are usually recorded for two hours and at 15 minute intervals. It is determined that within the study hour, the volume varies within each 15 minute block and one of these 15 minutes block having the maximum flow rate. When this maximum 15 minutes flow rate is multiplied by four, the resulting volumes are called "design peak period volume" or "peak flow rate volume." This is illustrated in Table 4-3.

Table 4-3 Peak flow rate.

Count Period	Vehicles	Rate of Flow (vph)	Peak Rate Flow (vph)
7:00 - 7:15	35	140	140
7:15 - 7:30	28	112	
7:30 - 7:45	20	80	
7:45 - 8:00	15	60	

PHF is defined as the ratio of total hourly volume to the maximum 15-minute rate of flow within the hour.

Table 4-2 Average daily traffic volume.

Location Gary between Holt Avenue and Third Street

AM Period	NB	Hourly Total	SB	Hourly Total	Both Direction	PM Period	NB	Hourly Total	SB	Hourly Total	Both Direction
12:00	52		24			12:00	240		144		
12:15	61		27			12:15	210		157		
12:30	39		23			12:30	216		214		
12:45	23	175	51	125	300	12:45	193	859	243	758	1617
1:00	31		23			1:00	210		220		
1:15	24		41			1:15	216		169		
1:30	26		18			1:30	174		216		
1:45	31	112	26	108	220	1:45	214	814	198	803	1617
2:00	26		23			2:00	201		206		
2:15	11		38			2:15	176		243		
2:30	20		25			2:30	286		249		
2:45	33	90	29	115	205	2:45	259	922	301	999	1921
3:00	14		37			3:00	279		291		
3:15	30		26			3:15	225		210		
3:30	19		41			3:30	190		249		
3:45	13	76	30	134	210	3:45	210	904	206	956	1860
4:00	11		42			4:00	240		240		
4:15	16		68			4:15	216		253		
4:30	14		45			4:30	304		329		
4:45	26	67	62	217	284	4:45	264	1024	261	1083	2107
5:00	35		35			5:00	240		277		
5:15	26		43			5:15	194		264		
5:30	53		62			5:30	150		276		
5:45	26	140	48	188	328	5:45	198	782	214	1031	1813
6:00	24		51			6:00	216		246		
6:15	39		39			6:15	181		273		
6:30	68		65			6:30	220		201		
6:45	75	206	73	228	434	6:45	174	791	169	889	1680
7:00	71		59			7:00	187		150		
7:15	50		50			7:15	171		111		
7:30	87		87			7:30	175		139		
7:45	103	311	96	292	603	7:45	159	692	94	494	1186
8:00	160		114			8:00	138		99		
8:15	150		120			8:15	120		117		
8:30	125		127			8:30	98		93		
8:45	120	555	145	506	1061	8:45	83	439	107	416	855
9:00	127		167			9:00	120		55		
9:15	150		181			9:15	107		74		
9:30	165		159			9:30	128		40		
9:45	188	630	170	677	1307	9:45	90	445	89	258	703
10:00	211		177			10:00	99		118		
10:15	170		150			10:15	78		81		
10:30	154		170			10:30	45		98		
10:45	205	740	172	669	1409	10:45	77	299	65	362	661
11:00	170		211			11:00	56		81		
11:15	194		200			11:15	92		53		
11:30	241		230			11:30	68		74		
11:45	171	776	253	894	1670	11:45	43	259	65	273	532
Total Volume	3878	3878	4153	4153	8031		8230	8230	8322	8322	16552
Daily Total											24583

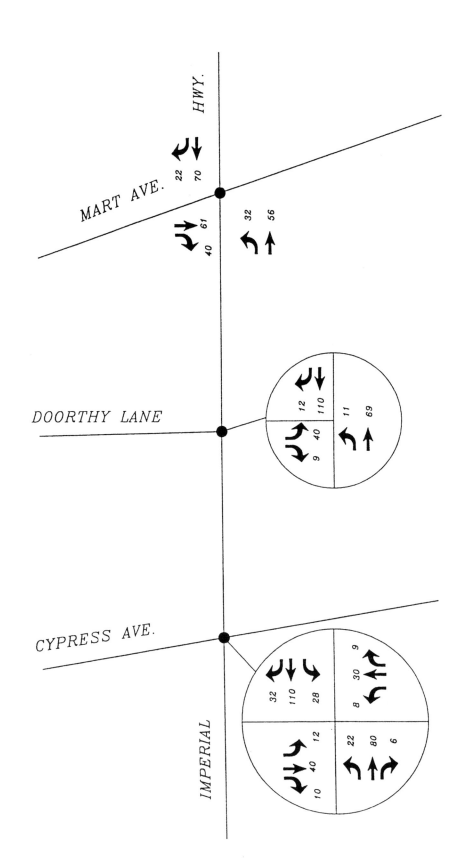

12 – VEHICLES PER HOUR (VPH)

Figure 4-3 Representation of turning movement counts.

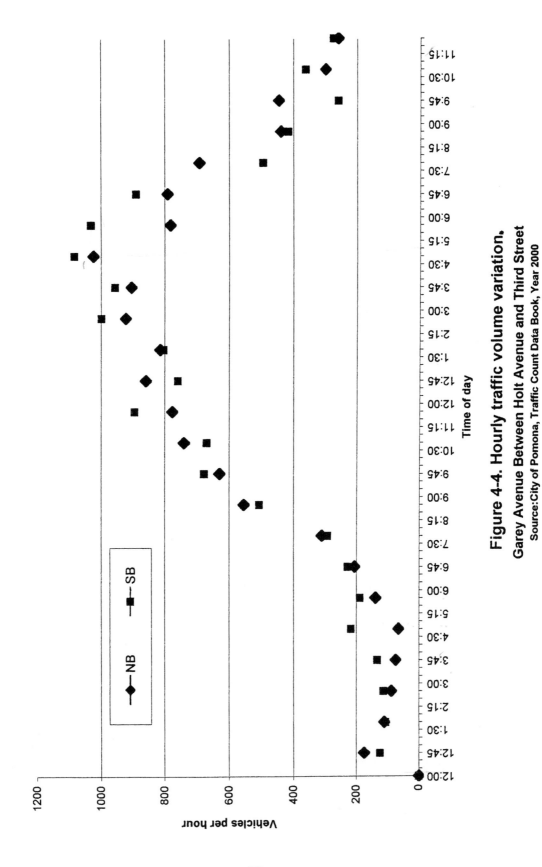

Figure 4-4. Hourly traffic volume variation,
Garey Avenue Between Holt Avenue and Third Street
Source:City of Pomona, Traffic Count Data Book, Year 2000

$$\text{PHF} \quad = \quad \frac{\text{Total Hourly Volume}}{\text{Peak 15 minute volume} * 4} \qquad \text{------- (4-1)}$$

Using the field data collected the peak flow rate and PHF for all turning movements can be calculated.

STUDY COMPONENTS

The main study components are to record the TMC's and calculate PHF.

Turning Movement Counts

In recording manual turning movement counts, proper procedure is essential to tally the vehicles traveling in each lane and direction. It is recommended that the intersection to be studied is visited 15 minutes before the scheduled period for collecting the data. The early site visit is required to familiarize with the study area and the traffic flow pattern at the intersection. Also, check the equipment used for recording TMC's. It is important to note the north direction before recording the data. The TMC's can be collected either using Manual or Electronic Count Boards. In using the manual count boards the data has to be transferred to the work sheet manually during every 15 minutes interval. If Electronic Count Boards is used, the data from the field can be directly dumped (transferred) into a computer. There are several sophisticated TMC electronic count boards available in the market today.

FORMULATING CONCLUSIONS AND RECOMMENDATIONS

1) Determine the peak rate of flow and PHF during both peak periods and off-peak periods for data recorded in the field.
2) Draw a sketch showing the intersection geometric and describe the intersection characteristics based on recorded TMC's.
3) What was the percentage composition of vehicles (cars, trucks, and buses) at the study intersection?
4) What is the percentage difference in total volume at the intersection between off-peak hour and peak hour TMC's?

WORK SHEET: LAB 4

R = RIGHT TURN
T = THROUGH
L = LEFT TURN

SHEET _____ OF _____
PEAK HOURS _____

MANUAL TRAFFIC COUNT SHEET

INTERSECTION OF _____ DATE _____ DAY _____

CITY _____ OBSERVER _____

TIME BEGINS	ON _____ DIRECTION			ON _____ DIRECTION			ON _____ DIRECTION			ON _____ DIRECTION		
	T	L	R	T	L	R	T	L	R	T	L	R

NOTES: U-TURNS TO BE COUNTED AS LEFT TURN.
TRAFFIC COUNTS INCLUDE MOTORCYCLES, BUSES, AND TRUCKS.

SOURCE: MOHLE, GROVER & ASSOCIATES

25

Lab 5

Measurement of Intersection Delay

LAB OBJECTIVES

The main objective of this lab is to determine the intersection delay values at a signalized intersection during peak period.

NEED FOR THE STUDY

Delay studies are conducted to evaluate the performance of the system. Also, these studies are conducted to evaluate certain elements within the system, such as traffic control devices (signals), stop signs, and any recent changes in the roadway links. For example, travel time delay studies (Lab 1) are conducted to determine the total time lost between an origin and a destination using alternate routes.

Similarly, the intersection delay study is conducted to evaluate the stopped-time delay for vehicles and pedestrians using the intersection. The determination of the total delay for all vehicles aids in mitigating the intersection to operate efficiently for both vehicles and pedestrians. The mitigation measures may include evaluation of existing traffic controls, checking the existing signal timing, and phasing.

GENERAL OVERVIEW

Delay is an integral part of travel. In general, delay can be defined as the additional time required or time lost to travel between an origin and destination. Delay is usually measured in terms of minutes or seconds per vehicle but when delay is multiplied by the total volume of traffic on a section of road, or at an intersection, will be in hours. There are different types of delays, as defined under Lab 1. They are fixed delay, variable delay, and travel time delay.

As expected, the intersection delays will be higher during peak periods and low during off-peak periods. Therefore, intersection delays are usually conducted during peak periods, depending on local conditions. If the volumes at the intersections are large (over 1,500 vehicles per hour), two or more persons will be used to record the traffic flow at the intersection. The intersection approach volume is recorded, usually at 15 seconds interval. If a vehicle is stopped during more than one time interval, the vehicle is counted during the remaining intervals as well.

STUDY COMPONENTS

The following components are studied in conducting intersection delay studies at an intersection. At any intersection, the vehicle could be either moving or stopped due to the existing traffic control device, delays at an intersection are defined and measured based on the state of the vehicle (moving or stopped).

Total Delay

The total delay caused for all vehicles stopped at an approach. An approach or leg of an intersection is defined as the direction from which vehicles are arriving at the intersection from various directions.

Average Delay per Vehicle

It is defined as the ratio of total delay for the approach (leg) divided by the total number of stopped vehicles.

Average Delay per Approach

It is defined as the ratio of total delay for the approach (leg) divided by the total number of vehicles in the approach, including both stopped and not-stopped vehicles.

Percentage of Vehicles Stopped

The number of vehicles stopped in the approach, divided by the total number of vehicles in the approach. The same could be determined for all approaches at an intersection.

FIELD WORK AND DATA COLLECTION

The following describes the manual method of collecting data for intersection delay study. The data collected include the recording of vehicles stopped and not stopped at an approach during specified interval of time. The time interval used ranges between 15 to 25 seconds usually depending on the cycle length of the signal at the intersection. The number of vehicles stopped at the approach during each observed interval is recorded.

If vehicles are stopped during more than one interval of recording, the vehicle is counted in all the interval periods during which the vehicle(s) remains in the study approach.
It is important to record separately the total volume passing the approach during the survey period. That is record the total number of vehicles actually stopping and those not-stopping. The recommended sample size is between 250 and 300 vehicles or when significant volume change is observed during survey period.

FORMULATING CONCLUSIONS AND RECOMMENDATIONS

1) What are the existing geometric and intersection characteristics of the study intersection (number of lanes, bus stops and pedestrian control)?
2) What is the total approach volume during study period?
3) From field survey data calculate the total delay, average delay and percentage of vehicles stopped.
4) What conclusions can be made about the study intersection? What improvements are required to reduce recorded delays?

WORK SHEET: LAB 5

MEASUREMENT OF INTERSECTION DELAY

DATE _____ DAY _____ WEATHER _____ PAGE _____

STUDENT NAME(S)/GROUP NUMBER _____

STARTING TIME _____ ENDING TIME _____

INTERSECTION _____ PEAK PERIODS _____

N/S STREET _____

E/W STREET _____

MOVEMENT OBSERVED _____

STARTING TIME	TIME INTERVAL USED IN RECORDING DATA				TOTAL VEHICLES	
	SEC	SEC	SEC	SEC	STOPPED	NOT STOPPED
TOTAL						

TOTAL VEHICLES STOPPED DURING SURVEY PERIOD _____

PERCENT STOPPING DATA PER APPROACH			
DIRECTION	VEHICLES STOPPED	VEHICLES NOT STOPPED	PERCENT OF VEHICLES STOPPED
EASTBOUND			
SOUTHBOUND			
WESTBOUND			
NORTHBOUND			

Lab 6

Sight Distance and Gap Study at Intersections

STUDY OBJECTIVES

The main objectives of this lab are to determine the minimum sight distance (based on posted speed on the major street) and to record available gaps in the major street traffic stream for vehicles exiting an unsignalized driveway or intersection during peak periods.

NEED FOR THE STUDY

One of the most important elements in designing a road section is the sight distance. Sight Distance is the length of roadway that is clearly visible to the driver without any obstruction. Suitable sight distance is required as the drivers on the road have different experience and training. Also, adequate stopping sight distance is to be provided to avoid accidents in designing intersections and driveways. A gap study is required to determine whether adequate gaps are available for the minor street traffic to enter the major street. The minor street can also be a driveway.

GENERAL OVERVIEW

Sight Distance

The American Association of State Highway and Transportation Officials (AASHTO) book on Geometric Design of Highways and Streets discusses sight distance in four steps:

1) The distance required for stopping, applicable on all highways
2) The distance required for the passing vehicles, applicable only on two- lane highways
3) The distances needed for decisions at complex locations
4) The criteria for measuring these distances for use in design

The stopping sight distance to be provided has two major components. The first component relates to the distance traveled by the vehicle after sighting the obstacle. The second component is the distance traveled after applying the brakes to stop the vehicle. These two components in general are referred to as brake reaction distance and braking distance, respectively.

The minimum reaction time for approximately 90 percent of the drivers is estimated as 2.5 seconds. This time is not completely adequate for complex situations such as blind intersections, but could be adopted for normal hazards, such as conflicts with parked vehicles, and exiting vehicles from driveways.

30

Gap Study

A driver at a stop sign, yield sign, or driveway (exiting) about to make a turn (left/right/through) will execute the desired movement only when adequate "gaps" are available in the traffic flow. This situation usually occurs to drivers from minor streets and also at driveways along major streets that are unsignalized. The avail able gaps on major streets depend of time of day and volume of traffic. There are two major elements that need to be considered. They are:

- Gaps available in major street traffic flow
- Acceptance of available gaps by the minor street driver

One of the most important elements for the major street driver is the sight distance. Assuming that the minor street driver has committed an error, the major street driver must be able to stop the vehicle or reduce speed to a safety level. Thus, in permitting turning movements from minor streets and driveways along major streets, it is necessary to check for minimum stopping sight distance and check for availability of acceptable gaps.

The gap acceptance by the minor street driver depends on the number of lanes on the major street, speed, volume, sight distance, driver characteristics (experience and training), and type of vehicle. The gaps available are recorded in seconds.

STUDY COMPONENTS

The breaking distance for a vehicle can be estimated using the following equation:

$$d \quad = \quad \frac{V^2}{30_g} \quad \text{--------- (6-1)}$$

Where,

$$d \quad = \quad \text{braking distance, feet.}$$
$$V \quad = \quad \text{initial speed, Miles per Hour.}$$
$$f \quad = \quad \text{coefficient of friction.}$$

The coefficient of friction "f" varies, depending on the pavement condition (wet or dry). Lower coefficient values are applicable on wet surfaces. Also, it depends on type of vehicle, type and condition of pavement surface, and air pressure of tires. Also, it depends on the braking capability of the vehicle and speed. For design purposes, the minimum sight distance is defined as the total distance traveled during the brake reaction time and the distance required stopping the vehicle to a stand still. The effect of upgrade and downgrade is to decrease and increase safe stopping sight distance, respectively. The following is the standard formula used to determine braking distance on grades:

$$d = \frac{V^2}{30(f \pm g)} \quad \text{------------- (6-2)}$$

Where 'g' is the percent of grade divided by 100 and all other terms are the same as defined under equation 6-2.

FIELD WORK AND DATA COLLECTION

For conducting gap study, the required minimum sight distance is calculated based on posted speed of the major street using equations 6-1 or 6-2. The calculated sight distance is marked on either side of the driveway or intersection on the major street as shown in Figure 6-1 from the centerline. The transverse markings are made using tapes on the roadway. A gap is defined as the period in which there are no vehicles within the marked region (Figure 6-1). That is, when the exiting driver(s) can make a safe turning movement onto the major street. The gaps available are recorded during peak hour of the major street to determine the safety and delay for the minor street traffic. The gaps are checked for 30 to 60 minutes at a site.

FORMULATING CONCLUSIONS AND RECOMMENDATIONS

1) What are the Major Street and Minor Street characteristics?
2) What is the estimated minimum sight distance based on posted speed of the major street?
3) What is the total gap in seconds available for the minor street traffic? What is the main cause of these gaps?
4) In your opinion is it safe to make turning movements from the minor street during peak periods and why?
5) What are the main conclusions from the study?
6) What design recommendations do you make to improve the flow of traffic from the minor street and reduce potential accidents?

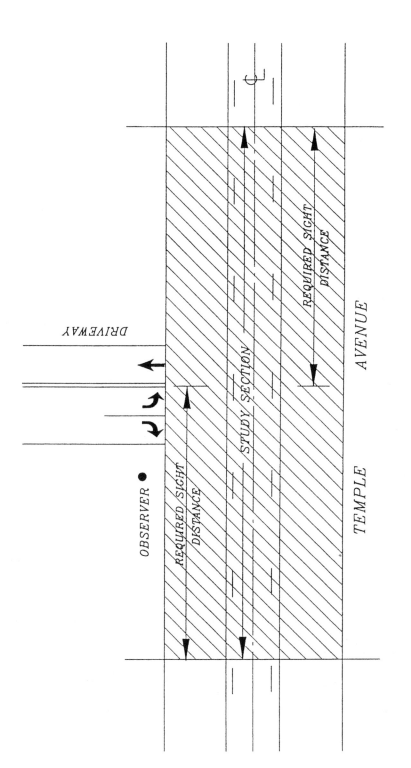

Figure 6-1 Sight distance and gap study.

LEGEND

STUDY SECTION (UNSIGNALIZED INTERSECTION)

WORK SHEET: LAB 6

SIGHT DISTANCE AND GAP STUDY

DATE _____ DAY _____ WEATHER _____ PAGE _____

STUDENT NAME(S)/GROUP NUMBER _____

STREET NAME _____

POSTED SPEED _____ MILES PER HOUR

REQUIRED SIGHT DISTANCE _____ FEET

TIME OF SURVEY (OFF-PEAK HOUR) _____

TOTAL NUMBER OF VEHICLES OBSERVED _____

STARTING TIME _____ ENDING TIME _____

STARTING TIME	ENDING TIME	NUMBER OF GAPS IN SEC.	TOTAL GAPS IN SEC.	CUMULATIVE GAPS	COMMENTS
EXAMPLE 8:00:00	8:02:00	8, 6, 14, 20	48	48	LEFT TURN 2 VEH.

TOTAL GAPS IN SECONDS = _____

Lab 7

Saturation Flow Rates

STUDY OBJECTIVES

The main objectives of this lab are to field measure the saturation flow rates and loss time at a signalized intersection during peak period.

NEED FOR THE STUDY

In calculating the Level of Service (LOS) of a signalized intersection, saturation flow rate plays a key role in determining the capacity. As per the 1995 Highway Capacity Manual (HCM), the capacity of a signalized intersection is based on saturation flow rate. The LOS in general indicates the operational qualities of an intersection and it is usually represented alphabetically from A to F. LOS A represents delay less than 5 seconds and LOS F represents delay more than 60 seconds at the intersection for all vehicles.

Saturation flow rate and loss time are two critical items in any signal timing analysis. Loss time is the time lost in the very beginning of the green time, when vehicles are starting to depart from the approach lane(s). These two important elements form the basis of delay methodology adopted by 1995 HCM.

GENERAL OVERVIEW

Saturation flow rate is the maximum rate of flow (vehicles) that can pass through an intersection approach or lane group under existing roadway and traffic conditions, assuming that the approach or lane group has 100 percent real time available as effective green time. Thus, saturation flow rate is expressed in units of vehicles per hour of green time (vehicles/hour of green, vphg).

In general, where saturation flow rate data is not available, an ideal value of 1900 vphg is used for through lanes, 1700 vphg for left, and 1600 vphg right turn lanes. The loss time used is usually two seconds in most cases. The value of saturation flow rate is affected by the following factors at an intersection:

1) Location of parking
2) Number of lane(s) in each group
3) Composition of heavy vehicles
4) Approach grade
5) Blocking effect of local transit
6) Area type (Central Business District)

7)   Lane group (through, left or right)

8)   Driver and road characteristics (type of vehicle, lane width, and pocket
        length)

Saturation flow rate is used in determining the capacity of a given lane group or approach at a signalized intersection. It is calculated by the following equation:

$$C_i = S_i * (g/c)_i \qquad \text{-------------} \quad (7\text{-}1)$$

Where: C_i =   Capacity of lane group or approach i (veh / hr)

            S_i =   Saturation flow rate for lane group or approach i
                            (veh / hr of green)

            $(g/c)_i$ =   Green ratio for lane group or approach I

STUDY COMPONENTS

The study components are to determine the saturation flow rate and loss time in the field. Also, to calculate the capacity of a lane group or approach using equation 7-1.

FIELD WORK AND DATA COLLECTION

The fieldwork and data collection is conducted as follows: Saturation flow rate measurement at a signalized intersection will require at least two persons to record the data. Figure 7-1 shows how to calculate loss time and saturation flow rate using the field data. In Figure 7-1, headway is defined as the time between two successive vehicles in a traffic lane as they pass a point (stop bar) on the roadway, measured from front bumper to front bumper, in seconds.

The following steps are to be observed in collecting the field data for saturation flow:

Complete intersection data work sheet and visit site 15 minutes before data recording. Record phasing sequence and cycle time of the signal. Select a suitable observation point depending on the approach selected (Stop Bar in Figure 7-1). Use the road markings as reference point or mark appropriate reference point on the road using a tape. The reference point will form the basis for all data readings. A minimum of ten cars should be in the queue is required for any data recording. Record the required data per work sheet. It is essential that the time recorder (person) coordinate work with data recorder in the field. If large number of heavy vehicles (trucks and buses) is present take readings. If loss time is to be determined, record the time of the first three cars as indicated in Figure 7-1.

FORMULATING CONCLUSIONS AND RECOMMENDATIONS

1) Draw sketches of the intersection geometric (Figure 4-1).
2) What is the signal phasing sequence and cycle time at the study intersection?
3) What is the loss time and saturation flow rate at the study intersections for each movement studied?
4) What factors influenced the saturation flow rate at the study intersection?
5) Is it possible to increase the saturation flow rates for any one movement(s) at the study intersections?
6) What improvements can be made to increase mobility and safety at the intersection?

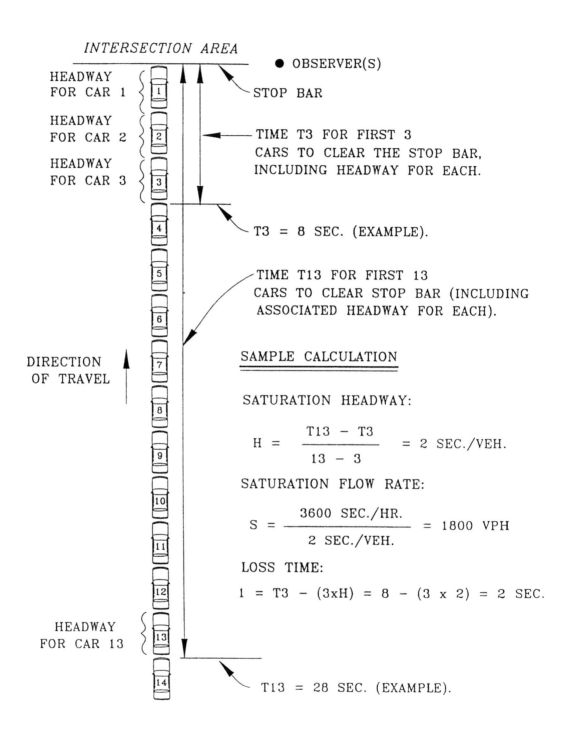

INTERSECTION AREA

● OBSERVER(S)

STOP BAR

HEADWAY FOR CAR 1

HEADWAY FOR CAR 2

HEADWAY FOR CAR 3

TIME T3 FOR FIRST 3 CARS TO CLEAR THE STOP BAR, INCLUDING HEADWAY FOR EACH.

T3 = 8 SEC. (EXAMPLE).

TIME T13 FOR FIRST 13 CARS TO CLEAR STOP BAR (INCLUDING ASSOCIATED HEADWAY FOR EACH).

DIRECTION OF TRAVEL

SAMPLE CALCULATION

SATURATION HEADWAY:

$$H = \frac{T13 - T3}{13 - 3} = 2 \text{ SEC./VEH.}$$

SATURATION FLOW RATE:

$$S = \frac{3600 \text{ SEC./HR.}}{2 \text{ SEC./VEH.}} = 1800 \text{ VPH}$$

LOSS TIME:

$$1 = T3 - (3xH) = 8 - (3 \times 2) = 2 \text{ SEC.}$$

HEADWAY FOR CAR 13

T13 = 28 SEC. (EXAMPLE).

Figure 7-1 Measurement of saturation flow rate.

WORK SHEET 'A': LAB 7

INTERSECTION DATA

LANE CONFIGURATION & GROUP

N/S STREET

DATE _____ DAY _____

LOCATION _____

INTERSECTION OF _____

OBSERVER(S) _____

WEATHER _____

E/W STREET

NUMBER OF LANES:	LEFT	THROUGH	RIGHT
NB	___	___	___
SB	___	___	___
EB	___	___	___
WB	___	___	___

LT+TH

TH+RT

S = SHARED LANE

RECORDED LANES WITH MORE THAN ONE MOVEMENT.

EXAMPLE: THROUGH WITH RIGHT TURN, THROUGH WITH LEFT TURN, ETC.

NAME	MAJOR STREET	MINOR STREET	COMMENTS
POSTED SPEED (MPH)			
CLASSIFICATION			
GRADE % ($\pm$)			
PEDESTRIAN MOVEMENT			
CONTROLS (S/UN/OT)			
% OF TRUCKS/BUSES			
CYCLE LENGTH (SEC)			
PARKING ON STREET			
TYPE OF INTERSECTION			
AREA TYPE (CBD)			
PHASING SEQUENCE			

S = SIGNALIZED   UN = UNSIGNALLED   OT = OTHER

WORK SHEET 'B': LAB 7

SATURATION FLOW RATE

DATE _____ DAY _____ WEATHER _____ PAGE _____

STUDENT NAME(S)/GROUP NUMBER _____

MOVEMENT OBSERVED _____

INTERSECTION _____

PEAK PERIOD _____ OBSERVER(S) _____

VEHICLES IN QUEUE	CYCLE 1			CYCLE 2			CYCLE 3			CYCLE 4			CYCLE 5			CYCLE 6		
	TIME	HV	T	TIME	HV	T	TIME	HV	T	TIME	HV	T	TIME	HV	T	TIME	HV	T

HV = HEAVY VEHICLES,   T = TURNING VEHICLES (LEFT/RIGHT)

Lab 8

Level of Service Analysis

LAB OBJECTIVES

The objectives of this lab are to determine the saturation flow rate, turning movement counts and level of service for an approach lane(s) at a signalized intersection during peak period.

NEED FOR THE STUDY

Level of Service (LOS), in general, is a qualitative measure describing operational conditions within a traffic stream. LOS is described in terms of such factors as speed and travel time, freedom to maneuver, traffic interruptions, comfort and convenience, and safety per the 1995 Highway Capacity Manual (HCM). The LOS at a signalized intersection is determined in terms of delay per HCM 1995. There are six LOS levels and they range from "A" to "F". The LOS per delay definition is as given in Table 8-1.

Table 8-1 Delay definition of level of service.

LOS	Stopped Delay for Vehicles (Seconds)
A	< 5.0
B	5.1 to 15.0
C	15.1 to 25.0
D	25.1 to 40.0
E	40.1 to 60.0
F	> 60.0

In the above table, LOS "A" indicates a very low delay and LOS "F" indicates a very large delay. LOS "F" is considered as not acceptable for all practical purposes. For most urban areas, a LOS "D" is acceptable and LOS "C" is the standard in most rural areas.

GENERAL OVERVIEW

In measuring LOS at a signalized intersection using the 1995 HCM delay methodology, the average stopped delay per vehicle is estimated using delay equation and compared with Table 8-1 to establish the existing LOS.

41

Delay is a measure of driver discomfort, frustration, fuel consumption, and lost travel time. Delay at an intersection depends on various factors, such as type of signal, volume of traffic, cycle length, progression and v/c (volume/capacity) ratio of each approach. The LOS in Table 8-1 has been established by the 1995 HCM based on the acceptability of various delays to the drivers. Therefore, it is important to note that the relation between capacity and LOS is not simple for all practical purposes.

The average stopped delay per vehicle at a signalized intersection for a given lane group is estimated using the following equation:

$$d = \frac{d_1 + d_2}{2} \quad \text{------- (8-1)}$$

$$d_1 = 0.38C[1-g/C] \; / \; \{1-(g/C)[\text{Min}(X,1.0)]\} \quad \text{-------- (8-2)}$$

$$d_2 = 173 X^2 \; \{ \, [(X-1) + [(X-1)^2 + (mX/c)]^{0.3} \, \} \quad \text{------- (8-3)}$$

Where;

d	=	stopped delay per vehicle for the lane group, sec/vehicle;
d_1	=	uniform delay, sec/veh;
d_2	=	incremental delay, sec/veh;
C	=	cycle length, sec;
g/C	=	green ratio for the lane group; ratio of effective green time to cycle length;
g	=	effective green time for lane group, sec:
X	=	v/c ratio for the lane group;
c	=	capacity of the lane group, vph; and
m	=	an incremental delay calibration term representing the effect of arrival type and degree of platooning.

The first term in the delay equation accounts for "uniform delay" assuming perfectly uniform arrivals and stable flow. Whereas the second term accounts for "incremental delay" due to the non-uniform arrivals and individual cycle failures. The delay value obtained in the above equation can be adjusted to reflect the assumed random arrival condition using a progression adjustment factor. For more details regarding the adjustments and definition of above terms, refer to the 1995 HCM Chapter 9 on signalized intersections.

Saturation Flow Rates

The capacity (c) in the delay equation is determined using equation 7-1 (Lab 7). Saturation flow rate is defined as the maximum rate of flow that can pass through an inter section approach or lane group under the existing roadway and traffic conditions, assuming that the approach or lane group has 100 percent of real time available as effective green time.

Saturation flow rate is expressed as vehicles per hour of green time. Saturation flow rate at a signalized intersection for a particular lane (through, left, or right) is determined per Lab 7. Table 8-2 shows the ideal saturation flow rates used for each type of movement at a study intersection.

Table 8-2 Ideal saturation flow rates.

Movements	Saturation Flow Rates
Through Lane	1900
Left Lane	1700*
Right Lane	1600*

*The reduction in saturation flow rate is attributed to adjustments required for turning movements. For definition of cycle length, green ratio, effective green time, types of random pat tern of arrivals and capacity, see "Glossary of Terms"

STUDY COMPONENTS

1) To record the existing traffic volume at the study approach lane.
2) To determine the saturation flow rate of study approach lane.
3) To estimate the delay and determine the LOS per 1995 HCM.

FIELD WORK AND DATA COLLECTION

The traffic volume (vehicles per hour) and saturation flow rate of each study approach lane(s) is recorded per Labs 4 and 7. The general instructions are to visit the site at least 15 minutes earlier than the data recording time and familiarize with the study intersection. Using field data, the delay equation 5 is used to determine the delay. The LOS is determined using Table 8-1 depending on calculated delay in seconds. The data collected in this lab can be used to determine the LOS using available traffic engineering operational software such as CAPSSI, and HCS. For details see section of "Computer Software Applications in Transportation Engineering" given after Lab 12.

FORMULATING CONCLUSIONS AND RECOMMENDATIONS

1) Draw the geometrics of the study intersection (Figure 1) and measure the total number of vehicles per hour at each study approach?
3) What is the loss time and saturation flow rate for the study movement?
4) Calculate the delay LOS using 1995 HCM delay equation for study approach lane (s) using field data?
4) Can the LOS be improved at the study intersection? List the improvements (if any).

Lab 9

Application of Poisson Distribution

LAB OBJECTIVES

The main objectives of this lab are to determine the probability of how many vehicles will arrive at a left turn pocket at a signalized intersection during peak periods.

NEED FOR THE STUDY

Traffic flow at an intersection when observed over time often follows a set pattern of flow. It is observed that, vehicles arrive at the intersection at regular intervals and the inter-arrival pattern is uniformly distributed. The probability of this occurrence at the intersection can be explained using mathematical probability distribution theories. The probability distributions are used to analyze recorded traffic data and predict anticipated pattern of flow. Studying the collected data to fit into an already known probability distribution provides the data analysis. One such probability distribution is Poisson distribution.

GENERAL OVERVIEW

Poisson distribution can represent several traffic flow situations where the flow is light and exhibits random behavior. Such practical situations include calculating the probability of delay at intersections, queuing analysis at tolls and parking lots. The Poisson distribution has several interesting applications in traffic analysis. The most common queuing models assume inter-arrival times and service times obey exponential distribution. The arrival and service rate follow a Poisson distribution. The Poisson distribution is generally stated as:

$$P(x) = \frac{M^x e^{-m}}{x!} \qquad \text{---------} \quad (9\text{-}1)$$

Where "M" is the average value of the event x per observation and M is defined as:

$$M = \frac{\text{Total cars observed}}{\text{Total observed frequency}} \qquad \text{-----------} \quad (9\text{-}2)$$

44

STUDY COMPONENTS

For example, equation 9-1 can be used to determine the expected number of cars that arrive at a left turn pocket in unit length of time. It is possible to determine the average number of arrivals along street sections by recording the arrival pattern in unit time.

The spacing between vehicles is defined as the distance between the front bumpers of one car to the successive car. The spacing phenomenon is studied using Poisson distribution. Using the Chi- Square test one can test the goodness of fit of a set of field data. The estimation of the arrival pattern aids in designing the facility, such as the length of a left or right turn pocket at a busy signalized intersection. The other two important probability distributions that are generally used are Binomial and Normal distributions. The details of probability distributions are well documented in the given references.

FIELD WORK AND DATA COLLECTION

The fieldwork involves collecting traffic data per the work sheet provided for this lab work. Select a busy intersection having an exclusive left turning lane. Complete the intersection data sheet (see work sheet lab 7) and record the number of cars arriving during a selected interval (say 15, 20, or 30 seconds) and the frequency of arrival. Use the work sheet and complete the calculations using Poisson distribution and determine the probability of arrival.

FORMULATING CONCLUSIONS AND RECOMMENDATIONS

1) What is the total number of cars arrived during the survey period?
2) What is the average number of arrivals (M)?
3) Determine the probability of X number of cars arriving during a particular interval?
4) Determine theoretical frequency F_i [F_i = (total observed frequency) * $P(x_i)$].
5) What improvements are required at the study intersection for better traffic flow?

WORK SHEET: LAB 9

APPLICATION OF POISSON DISTRIBUTION

DATE _____ DAY _____ WEATHER _____ PAGE _____

STUDENT NAME(S)/GROUP NUMBER _____

STARTING TIME _____ ENDING TIME _____

INTERSECTION _____ PEAK PERIODS _____

N/S STREET _____

E/W STREET _____

MOVEMENT OBSERVED _____ OBSERVERS _____

NUMBER OF VEHICLES X_i	OBSERVED FREQUENCY F_i	TOTAL VEHICLES OBSERVED	PROBABILITY OF X_i, $P(X_i)$	THEORETICAL FREQUENCY F_i
COL. 1	COL. 2	COL. 3 = COL. 1 × COL. 2		
TOTAL				

Lab 10

Application of Queuing Analysis

LAB OBJECTIVES

To verify the storage capacity of an existing drive-thru facility at a fast food restaurant, bank teller and facilities where queuing occurs.

NEED FOR THE STUDY

In transportation, the storage area, or throat length, for a particular facility where drive-thru is possible needs to be determined before constructing facilities such as fast food restaurants, banks, and parking areas to avoid queuing. The length of storage area required should be estimated and located in an area such that the arriving vehicles for service do not block or impede the movement of traffic in the surrounding area. Queuing analysis can be used to analyze and solve various problems related to storage length, and number of servers at a facility. For more details on the characteristics of queuing, refer to the books given at the end of this chapter.

GENERAL OVERVIEW

The formation of a queue is not uncommon in everyday work. For example, one can observe people waiting in line at a fast food restaurant, and cars waiting in line to enter a parking area. Queuing is present at almost all places and in most cases it is unavoidable. The reason for queuing is simple; the demand is larger than the service provided at a particular facility. Queuing can occur with a single server or multiple of servers. Figure 10-1 shows examples of single and multiple channels serving facility. In general, each serving area or station is called as a "channel." If there is only one serving station available, it is called a single channel problem. If there is more than one serving station, it is called a multiple channel problem. Usually customers arrive in a random fashion at a service facility. If service is available immediately, the customer passes, if not, the customer joins the queue. Therefore, a queue is formed when service is not available for a customer. The customer will wait till the server provides the required service. It can be a single or multiple server channels as in Figure 10-1.

STUDY COMPONENTS

Storage Length

The length of storage required at a drive-through facility (fast food restaurant or bank) is designed based on the expected number of vehicles arriving in unit time. The design has to consider the service time required per customer. Also, it becomes necessary to check whether the design storage length provided serves all vehicles assuming different

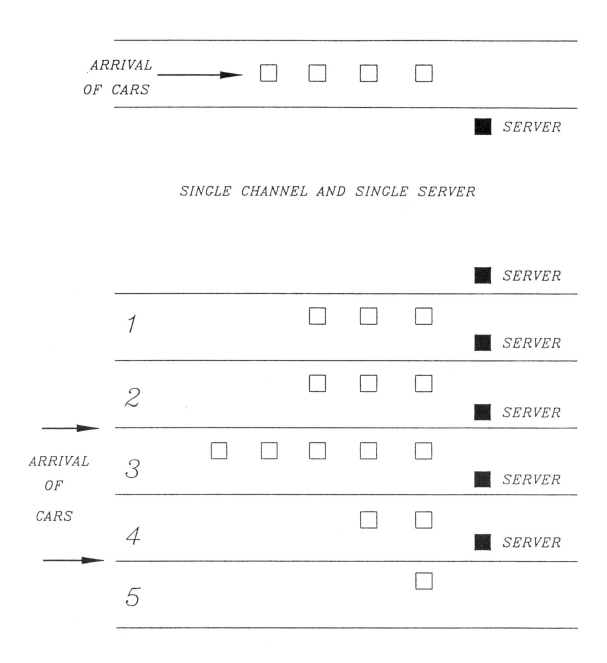

Figure 10-1 Single and multiple channels.

probability. For example, it will be necessary to determine if the throat length provided is able to accommodate the peak period traffic 95% of the time. Using the following equation 10-1 it is possible to solve the above type of queuing problems:

$$M = \frac{[\ln P(x > M) - \ln Q_m]}{[\ln p]} - 1 \qquad \text{---------- (10-1)}$$

Where:

M	=	Queue length which is exceeded p percent of the time
N	=	Number of service channels
Q	=	Service rate per channel (vehicles per hour)
P	=	$\dfrac{\text{Demand Rate}}{\text{Service Rate}} = \dfrac{q}{NQ} = $ Utilization factor

q is the number of vehicles arriving per hour

Q_m is obtained from Table 10-1, which provides the relationship between queue length, number of channels, and utilization factor. The above equation can be used to test whether a facility has adequate storage length or what is the required storage length to reduce delays. That is the equation can be used to determine if unacceptable delays are caused to vehicles in the queue at the facility by assuming suitable probability $P(x > M)$ in using the equation.

FIELD WORK AND DATA COLLECTION

The field work and data collection are conducted as follows:

- Select a drive-thru facility at a fast food restaurant or bank.
- Record the number of channels (service stations) at the facility.
- Record the number of vehicles arriving per hour (q) and the average service time.

Table 10-1 Table of Q_m values.

	N = 1	2	3	4	6	8	10
0.0	0.0000	0.0000	0.0000	0.0000			
0.1	.1000	.0182	.0037	.0008	.0000	0.0000	0.0000
0.2	.2000	.0666	.0247	.0096	.0015	.0002	.0000
.3	.3000	.1385	.0700	.0370	.0111	.0036	.0011
.4	.4000	.2286	.1411	.0907	.0400	.0185	.0088
.5	.5000	.3333	.2368	.1739	.0991	.0591	.0360
.6	.6000	.4501	.3548	.2870	.1965	.1395	.1013
.7	.7000	.5766	.4923	.4286	.3359	.2706	.2218
.8	.8000	.7111	.6472	.5964	.5178	.4576	.4093
.9	.9000	.8526	.8172	.7878	.7401	.7014	.6687
1.0	1.0000	1.0000	1.0000	1.0000	1.0000	1.0000	1.0000

FORMULATING CONCLUSIONS AND RECOMMENDATIONS

1)   Draw a sketch of the facility selected for data collection.
2)   What are the peak periods at the facility?
3)   Determine service rate, number of channels, and utilization factor for the facility.
4)   Determine the probable queue length for an assumed acceptable probability.
5)   What improvements are required to improve service at the facility?

WORK SHEET: LAB 10

APPLICATION OF QUEUING THEORY

DATE _____ DAY _____ WEATHER _____ PAGE _____

STUDENT NAME(S)/GROUP NUMBER _____

TYPE OF FACILITY _____

NUMBER OF CHANNEL(S) _____

NUMBER OF SERVER(S) _____

PEAK PERIODS _____

TOTAL NUMBER OF VEHICLES OBSERVED _____

OBSERVER(S) _____

STARTING TIME	ENDING TIME	NUMBER OF VEHICLES	TOTAL SERVICE TIME	COMMENTS

Lab 11

Parking Study

STUDY OBJECTIVES

The main objectives of this lab are to identify the various elements that relate to a parking study by examining an existing parking area.

NEED FOR THE STUDY

Parking studies are conducted to determine the demand at an existing or new parking area. Also, to check the parking facility for compliance with the local jurisdiction zoning codes. The basic design elements studied are the parking turnover, parking accumulation, vehicle occupancy rate, pavement markings, and traffic control devices. Also, parking studies are conducted to study the existing parking regulations and safety aspects of the parking area. Recommendations are made relative to proposed improvements to the existing conditions on-site. Parking studies are conducted to answer one or more of the following questions:

- How many parking spaces are required based on estimated trips (trip generation) to the activity or land use?
- How much land is allocated for parking and what alternatives are available?
- What is the cost of construction and maintenance?
- What kind of security, safety, regulatory compliance, financing and permits are required?
- What changes do you recommend for the access points?
- Who will pay for the operation and maintenance?
- How much will it cost to construct and operate the facility?

GENERAL OVERVIEW

Parking is an integral part of the transportation system. Everyone who owns an automobile expects to find parking at their origin or destination. The origin could be an apartment; the destination could be a university. Thus, it is important to structure the parking requirement based on anticipated demand. The number of parking spaces provided at a land use usually depends on the type of activity. For example, the number of parking spaces required for an apartment is at least 1-2 spaces per unit. For a shopping center, the parking requirement is estimated based on the gross floor area of construction; for example, 4-5 spaces per 1,000 square feet of gross floor area. While the standards vary among jurisdictions, however parking requirements are generally based on land use. Parking requires large land area, but the prime land available for parking is limited. This indicates that parking and land use are inter- related, and thus parking is a very important element to be considered in land use planning and zoning.

Local zoning ordinances are developed to specify the required parking spaces for each type of land use. Also, zoning ordinances of each jurisdiction will give complete information with respect to the minimum acceptable parking geometric, setbacks, and other items related to parking. These ordinances can be obtained from the local juris dictions.

STUDY COMPONENTS

The following parking elements are studied in conducting a parking study:

Parking Area Characteristics

The land use surrounding the site, the size of the existing parking area, access locations, approach road characteristics, general stall geometric and lot lay out are studied in detail. The existing number of parking spaces for visitors, emergency vehicles, reserved parking, motorcycles, bicycles, and other special parking spaces are also recorded.

Parking Demand and Supply

The parking demand is the number of vehicles coming to a particular land use area during the operating hours. When the rate of arrival during specific time periods in the day is compared to the available number of parking spaces will determine the parking requirements for that development. Similarly, parking supply is the number of parking spaces to be provided based on the increase in demand.

In determining parking demand, the available parking spaces are recorded. The turn over and accumulation pattern is surveyed during peak periods and off peak periods. The parking turnover is recording the number of times the same parking space is used by various users of the parking lot during the day. Parking turnover mainly depends on the type of land use and the duration of parking required by each vehicle. Parking accumulation is the number of vehicles parked at any given time.

The vehicle occupancy rate is determined at all access points to the parking area at large parking lots. The vehicle occupancy rate is defined as the average number of persons per vehicle entering the parking area. Vehicle occupancy studies are usually conduct for parking spaces used for special events such as sports arena, convention center and cinemas. The vehicle occupancy survey is used to estimate the actual number of vehicles arriving in order to obtain parking variance (reduction in parking spaces) from zoning ordinances.

The percentage composition of different vehicles on site (large, medium, and small) on site is recorded. Also, recorded are other types of vehicles in the parking area such as buses, delivery vans and other special types of vehicles. The queuing characteristics at entrance booths and access points are studied for possible conflicts in traffic circulation on-site and off-site.

Parking Control

A study of existing general regulations, traffic control devices, fee structure and general signs posted is conducted.

Parking Security

The parking security study is necessary to check the types of access control (gated or open access), lighting on site, and restricted parking areas.

FIELD WORK AND DATA COLLECTION

A visit to the site is required to collect the data for a parking study. Also, parking layout maps or plans if available can be used to record some of the above data. The surrounding land use type is recorded, as it is required to estimate the demand on the parking area. During preliminary data collection, it is necessary to collect zoning ordinances and local jurisdiction requirements. The geometric of the approach road and driveway are recorded as shown in Figure 4-1.

The design of the access points determines the ease with which vehicles can enter and exit the parking lot without causing undue congestion or queuing on the access street. Figure 11-1 shows a general parking lot stall geometric and lot lay out. These two elements are important in parking design, as they determine the number and type of vehicles that can be allowed to park.

A parking lot usually provides parking spaces for special uses such as visitors, handicapped, emergency vehicles, delivery trucks, and motorcycles. It is necessary to record the number of spaces allocated for such uses. Recording the number of parking spaces available for different vehicles in the parking lot will help to determine the total number of vehicles that can be parked at any time.

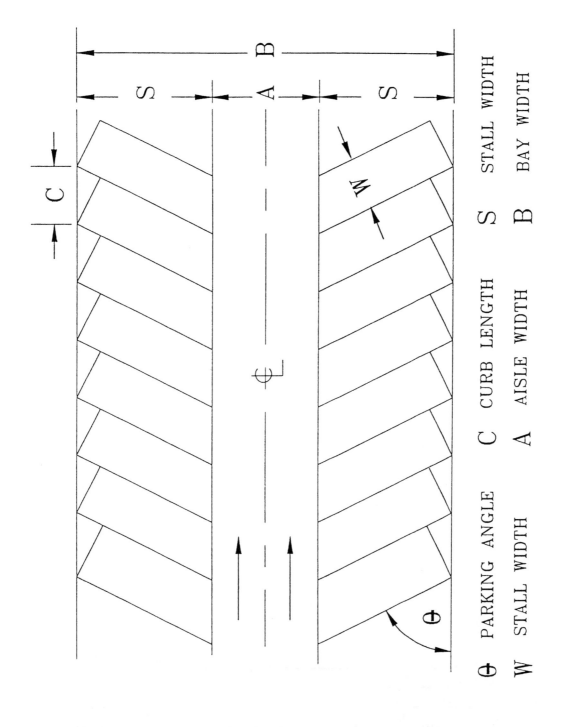

Θ   PARKING ANGLE   C   CURB LENGTH   S   STALL WIDTH

W   STALL WIDTH   A   AISLE WIDTH   B   BAY WIDTH

Figure 11-1   General parking stall geometrics.

The parking turnover is calculated by recording the number of vehicles using the same parking space during the day. Parking Accumulation is the number of vehicles in the facility at any given time. The turnover and accumulation studies are conducted during peak periods to determine the peak demand on the parking system. Also, it is necessary to record the length of time vehicles are parked in the facility. The vehicle occupancy rate is determined by conducting a survey at the parking area driveways or access points. The survey consists of recording the number of persons in each vehicle entering the facility.

The vehicle occupancy rate is the ratio of the total number of persons divided by the total number of vehicles entering the parking area during the survey hour. Table 11-1 shows sample vehicle occupancy rate survey data. The survey is conducted during peak periods for at least one to two hours duration.

The vehicle occupancy rate studies are mainly conducted at special event parking areas such as sporting stadiums, convention center, and cinemas. That is, at places where large number of parking spaces are to be provided and number of persons per car is expected to be greater than one.

In the above survey period the total number of vehicles observed is 10 and the total number of persons recorded is 20. Therefore, the vehicle occupancy rate for the above survey data is 2.0. The percentage composition of each type of vehicle in a parking lot is an important data required to design appropriate parking stall geometric. Additionally it is advantageous to provide parking areas for different types of vehicle separately and thus maximizing the area available for parking and circulation.

Table 11-1. Vehicle Occupancy rate survey.

Vehicle Number	Persons in Each Vehicle
1	3
2	1
3	1
4	3
5	2
6	4
7	1
8	1
9	2
10	2

The formation of queues at the access points may cause congestion on the main access road to the parking lot resulting in unnecessary delay. Therefore, it is necessary to study the queuing pattern (the number of vehicles entering the parking lot during peak periods) at the entrance and exit points of the parking lot. This type of study will help to determine the required storage or throat length at the entrance or exit driveways. It is possible to calculate the required storage or throat length using queuing analysis (Lab 10).

There are several general signs used in a parking area. For example, there are signs showing handicapped parking spaces, and special parking areas for buses, trucks, delivery vans and recreational vehicles. Also, general signs include signs showing regulations applicable to the parking area, directional signs to various land uses, and general instructions to the public using the parking area.

Providing parking security is an important element in designing a parking area. By providing gated access, lighting, and personnel to supervise the parking area parking security is achieved.

FORMULATING CONCLUSIONS AND RECOMMENDATIONS

1) Draw a general sketch of the study parking area, access points, and typical parking stall plan?
2) Does the study parking area and stall geometric meet the local zoning codes?
3) What are the parking turnover and accumulation characteristics for the parking area?
4) What does the vehicle occupancy rate survey tell you?
5) What is the land uses surrounding the parking lot?
6) Prepare a table showing percentage composition of cars.
7) List traffic signs and regulatory devices used in the parking facility.
8) Does adequate security exist at study parking area? If not, what improvements are required?

WORK SHEET: LAB 11

VEHICLE OCCUPANCY RATE (VOR)

DATE _____ DAY _____ WEATHER _____ PAGE _____

STUDENT NAME(S)/GROUP NUMBER _____

PEAK PERIODS _____

LOCATION _____

STARTING TIME _____ ENDING TIME _____

VEHICLE NUMBER	NUMBER OF PERSONS	VEHICLE NUMBER	NUMBER OF PERSONS	VEHICLE NUMBER	NUMBER OF PERSONS

TOTAL NUMBER OF VEHICLES = _____ VOR =

TOTAL NUMBER OF PERSONS = _____

Lab 12

Pedestrians and Bicycle Study

LAB OBJECTIVES

The main objectives of this lab are to study an existing pedestrian and bicycle facilities at a large project site such as a Convention Center, Recreational Park, Shopping Center, or University Campus.

NEED FOR THE STUDY

The design of transportation facilities to accommodate pedestrians and bicycle users are very much essential to reduce the number of accidents involving them and the motor vehicles. Most pedestrians and bicycle accidents occur at intersections, parking areas, and along road sections. The total fatality of pedestrians in the U.S. was 8,400 in 1981, with 3,800 fatalities during the daytime and 4,600 fatalities during the nighttime.

The major pedestrian facility requirements are at street corners, particularly in urban areas. At signalized intersections, it is necessary to provide pedestrian actuated signals to allow pedestrians to cross the street safely. Also, in urban areas, sidewalks are required to be provided to accommodate pedestrians traversing between land uses.

Bicycle use has been on the increase and so has been the accident rate. Presently, bicycling is looked as an alternative mode of transportation. This trend is increasing and programs are required to provide proper bicycle facilities on streets and parking areas.

The design of facilities for pedestrians and bicyclists varies strictly due to the characteristics of these two users. Recently, transportation planning has taken a careful look at design elements required for both pedestrians and bicycle users.

GENERAL OVERVIEW

Pedestrian Facilities

The most common pedestrian facilities are the sidewalks, walkways, crosswalks, and street sections at intersections. The design of facilities for pedestrians is similar to the flow of traffic. This is due to the similarities in flow patterns between vehicles and pedestrians. The walkways are designed for maximum anticipated flow of pedestrians, such as 25 persons per minute per foot of width, moving at approximately 150 feet per minute. The actual capacity of a crosswalk is controlled by the characteristic of traffic flow on the road section and adopted local lanes with respect to pedestrian movement on the streets.

Bicycle Facilities

The use of bicycles has been classified under Human Powered Transportation. The normal speed of a bicycle ranges between seven to fifteen miles per hour. Sometimes it has been measured at 30 miles per hour. The separation observed between two cycles driven at 10 miles per hour was recorded as 2.5 feet between handlebars. The following dimensions of a bicycle are:

Handlebar's width	1.96 ft.
Length	5.75 ft.
Pedal clearance	0.50 ft.

The use of roads by bicyclists is much more than pedestrians. A bicycle has been defined as a vehicle having two tandem wheels propelled solely by human power. The bikeways have been classified as follows by the Federal Highway Administration (1980):

Class I bikeway: completely separated from vehicular traffic and within an independent right-of-way or the right-of-way of another facility. Bikeways separated from vehicles but shared by both bicycles and pedestrians.

Class II bikeway: pavement markings or barriers mark part of the roadway or shoulder. Vehicles parking, crossing, or turning movements are permitted within the bikeway.

Class III bikeway: shares' right-of-ways with motor vehicles and are designated by signing only. There is hardly any protection from motor vehicles, although the signing helps to make the motorist aware of the presence of bicyclists.

The *1995 Highway Capacity Manual (HCM)* classifies bicycle facilities in two basic forms. Bike lane is a portion of a roadway that is striped, signed, and marked for exclusive or preferential use of the bicyclists. A bike path is defined as a bikeway if physically separated from motorized vehicular traffic, either within the highway right-of-way or within an independent right-of-way.

STUDY COMPONENTS

Design of Pedestrian Facilities

In designing pedestrian facilities, it is assumed that pedestrians require 2.5 feet each (width) and for pedestrians walking together need 2.2 feet each. In designing pedestrian right-of-way at signalized intersections, a speed of 4 feet per second is assumed for a pedestrian to cross from curb-to-curb. In designing any pedestrian facility, it is required to consider comfort, convenience, safety, security, and economy. The Level of Service for a walkway ranging from "A" to "F" is given in Table 12-1.

Table 12-1 Pedestrian Level of Service for Walkways.

L.O.S	Space		Average	Volume/Capacity
	(ft/ped)	(ft/min)	(ped /min /ft)	(V/C)
A	130	260	2	0.08
B	40	250	7	0.28
C	24	40	10	0.40
D	15	225	15	0.60
E	6	150	25	1.00
F	< 6	< 150	Variable	Variable

Average condition for 15 minutes. Source: Highway Capacity Manual, 1995.

FIELD WORK AND DATA COLLECTION

Select a suitable site such as a Convention Center/Recreation Park/Shopping Center/University:

- Measure the width of the sidewalks and bicycle lanes.
- Record the number of pedestrians and bicycles using the facility.
- Compare the field data to determine the existing LOS.

FORMULATING CONCLUSIONS AND RECOMMENDATIONS

Pedestrian Study

1) What is the measured pedestrian flow rate per minute?
2) Sketch the geometric of pedestrian facility to scale.
3) What are the characteristics of study site?
4) What is your assessment of demand and adequacy of design elements?

Bicycle Study

1) What is the existing bicycle demand at site?
2) What type of facilities at site is provided for bicycles?
3) Draw a sketch of the bikeway geometric to scale.
4) Describe elements, such as average speed and stopping sight distance at site.
5) What is the observed impact of bike volume on motor vehicular traffic at site?
5) What improvements are required for pedestrian and bicycle facilities on-site?

Lab 13

Travel Demand Forecasting

STUDY OBJECTIVES

The main objective of this lab is to provide an introduction to the use, application and analytical basis of travel demand forecasting (TDF) process in transportation planning. The lab will provide the basics required to understand TDF process in transportation planning.

NEED FOR THE STUDY

Every day daily activities are conducted by going from one place (origin) to another place (destination) to complete tasks using different modes. The tasks could be like going to work, school, mall, hospital and university. The modes used to travel could be car, bus, rail, vanpool and bicycle. Figure 13-1 shows trip making patterns between various zones (origin-destination) during daily activities. The majority of the people drive alone to reach their destination; particularly it is prevalent in the urban and sub-urban areas. Most auto drivers trying to reach their destinations experience congestion; the congestion could be caused due to large number of vehicles on the road, construction and accidents. These delays are classified as recurring and non-recurring delays. So, it is necessary to understand current travel demand on the existing network. The future demand in an area may occur due to growth in the region and influx of new developments and population. Therefore a methodology is required to estimate the future travel demand based on current demand and anticipated growth in the area.

GENERAL OVERVIEW

Travel demand forecasting is a process that attempts to predict the future demand for travel within a region. The process is complex. This lab will provide the basic knowledge to understand TDF and its application as related to transportation planning. TDF is also called as urban travel demand forecasting. It is a tool developed to support the urban transportation planning process. The output from modeling process assists Planners, decision-makers and the public to develop feasible alternatives to alleviate future demand on the system. The model results provide future number of cars (autos) and other modes using the roadway network. The output may include volume of passengers using the transit system (rail and bus), the demand for high occupancy vehicle (HOV) lanes and toll routes.

63

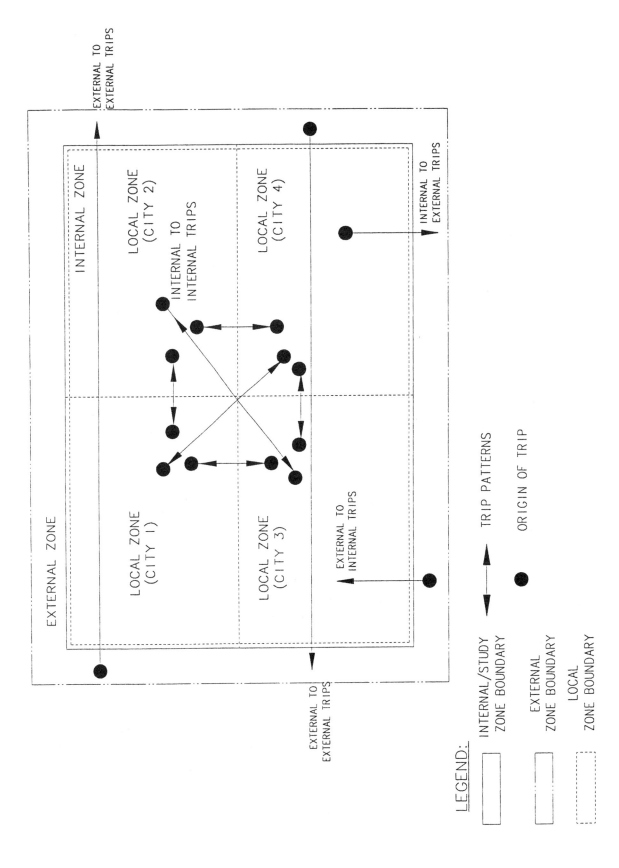

Figure 13-1. Types of trip patterns in daily activities.

LEGEND:

INTERNAL/STUDY
ZONE BOUNDARY

EXTERNAL
ZONE BOUNDARY

LOCAL
ZONE BOUNDARY

TRIP PATTERNS

ORIGIN OF TRIP

D:\DR-MURTHY\FIG-2.DWG   9 16 1999

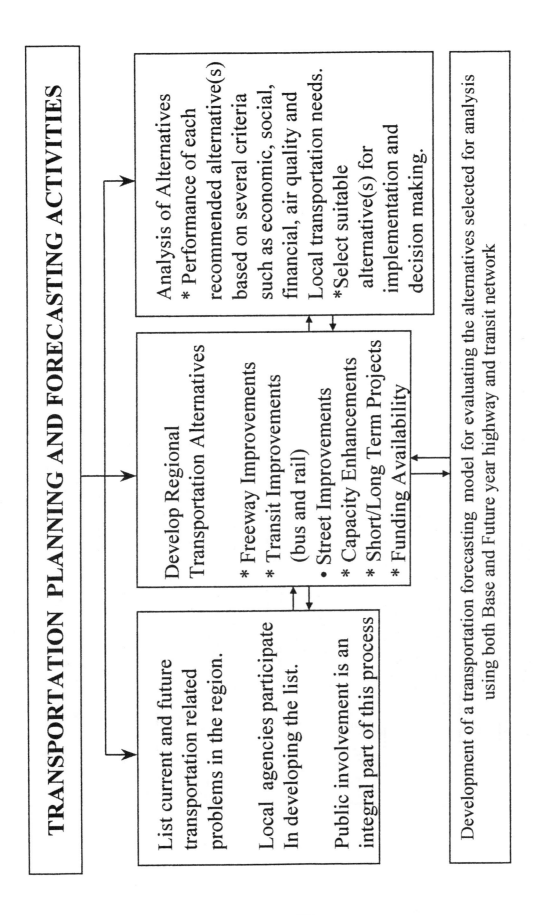

TRANSPORTATION PLANNING AND FORECASTING ACTIVITIES

List current and future
transportation related
problems in the region.

Local agencies participate
In developing the list.

Public involvement is an
integral part of this process

Develop Regional
Transportation Alternatives

* Freeway Improvements
* Transit Improvements
  (bus and rail)
• Street Improvements
* Capacity Enhancements
* Short/Long Term Projects
* Funding Availability

Analysis of Alternatives
* Performance of each
recommended alternative(s)
based on several criteria
such as economic, social,
financial, air quality and
Local transportation needs.
*Select suitable
alternative(s) for
implementation and
decision making.

Development of a transportation forecasting model for evaluating the alternatives selected for analysis
using both Base and Future year highway and transit network

Figure 13-2 Transportation planning process.

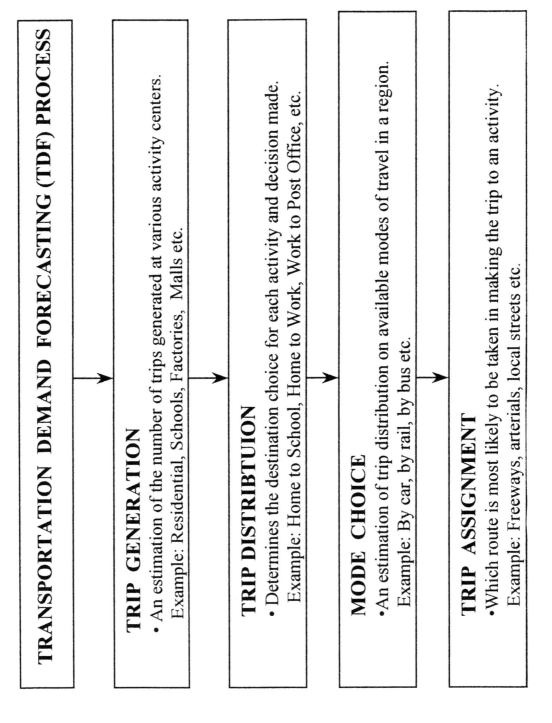

TRANSPORTATION DEMAND FORECASTING (TDF) PROCESS

TRIP GENERATION
• An estimation of the number of trips generated at various activity centers.
Example: Residential, Schools, Factories, Malls etc.

TRIP DISTRIBTUION
• Determines the destination choice for each activity and decision made.
Example: Home to School, Home to Work, Work to Post Office, etc.

MODE CHOICE
•An estimation of trip distribution on available modes of travel in a region.
Example: By car, by rail, by bus etc.

TRIP ASSIGNMENT
•Which route is most likely to be taken in making the trip to an activity.
Example: Freeways, arterials, local streets etc.

Figure 13-3 Basics of transportation demand forecasting process.

66

Figure 13-2 shows the basics of transportation planning and forecasting activities. At some point in time the traffic on the roadway network exceeds existing capacity creating recurring congestion leading to delays. The question is how to accommodate future traffic in the region. Several agencies in various jurisdictions meet and discuss about alternatives. In order to test these alternatives travel demand models are built for the current and future years.

Figure 13-3 shows the actual steps involved in developing travel demand model. A four-step process is used in travel demand forecasting. The four- steps are:

1. Trip Generation.
2. Trip Distribution
3. Mode Choice
4. Trip Assignment.

STUDY COMPONENTS

Trip generation is the first step in TDF process. A trip generation model estimates the number of trips generated or attracted to a specific zone such as residential, school, shopping mall, industrial area or activity centers such as sporting events, museums and theme parks. In modeling process the area in a region is divided into traffic analysis zones (TAZ's). A trip generation model estimates the number of persons or vehicle trips to and from each TAZ. The trips can be estimated based on number of persons, number of autos, number of employees or on an aggregate at the zone level.

The data for trip generation is collected through home surveys, activity surveys (trip dairy), workplace survey and other types of data collection process. For modeling purposes the major trips are classified based on trip purposes as,

1. Home Based Work (HBW); going from home to work
2. Home Based-Other (HBO); going from home to school, shopping etc. and
3. Non-Home Based (NHB); going from office to shopping, post-office, meeting etc.

Also, trip origin (from) and destinations (to) are replaced by trip production and trip attraction for modeling purposes. Trip production and attraction models are developed using regression equations and other mathematical models for each TAZ. Sometimes it will be required to develop special generator models for large trip generators such as hospitals, large shopping malls, sports centers, airports and universities. Table 13-1 shows trip generation rates for a HBW trips.

Trip distribution is the next step in TDF process to determine where the trips are going? Trip distribution links the trip production in a zone with another zone to which the trips are attracted. The output from a trip distribution model is trip length, trip destination and traffic and passenger volume. For trip distribution the most generally used model is Gravity Model. This gravity model is based on the original Newton's gravitational concept put forward in

1686. The gravity model has been the most widely used trip distribution model to distribute traffic in an urban area.

Table 13-1 Trip generation rates.

	Basic	Retail	Service	Attractions per Household
Area Type	Trips per day or attraction	Trips per day or attraction	Trips per Day or attraction	
Central Business District	1.11	1.56	1.12	0.12
Urban Fringe	1.68	1.2	1.15	0.08
Urban Residential	1.68	0.99	1.49	0.09
Suburban Area	1.3	0.99	1.49	0.11
Rural Area	1.3	1.26	1.16	0.1

Source: Introduction to Urban Travel Demand Forecasting, FHWA, July 1998.

Gravity model used for trip distribution states that "trip interchanges between zones (TAZ's) is directly proportional to the relative attraction of each of the zones and inversely proportional to some function of the spatial separation between the zones."

$$T_{IJ} = \frac{P_I A_J}{\sum\limits_{J=1}^{n} A_J}$$

Where,

TIJ= Total trips produced in zone I and attracted to zone J

PI = Trips produced in zone I

AJ = Trips attracted in zone J

The gravity model has gone through some revisions in adopting it for trip distribution. The major changes were to add new parameters to express the effect of spatial separation (F_{IJ}) and socioeconomic characteristics of zones (K_{IJ}) that influence trip.

The revised gravity model is as follows:

$$T_{IJ} = \dfrac{P_I \mathbf{A}_J F_{IJ} K_{IJ}}{\sum\limits_{J=1}^{n} A_J F_{IJ} K_{IJ}}$$

Where,

F_{IJ} = Empirically derived travel time factor which expresses the spatial separation on trip-interchanges between zones.

K_{IJ} = It is a specific zone-to-zone adjustment factor which could allow for differences in social or economic linkages between the zones.

T_{IJ}, P_I and A_J are as defined above.

Friction factors are calibrated so that the number and lengths of trips from trip distribution model are close to those observed. The socioeconomic factors are applied to gravity model as a fraction; it is used sparingly only if friction factors cannot correct the model output.

Also, another important output from a trip distribution model is the trip length for each trip type which is later converted into a trip length frequency diagram (TLFD) as shown in Figure 13-4. The model TLFD is compared with observed TLFD. In Figure 13-4 the thick line shows the observed data and the dotted lines represent the model output. The model output closely follows the observed data. The model is further calibrated by adjusting the friction factors to duplicate observed data.

The third step in TDF is to find out which mode of travel is being used by commuters to reach their destinations. For this purpose a mode choice model is used. The mode choice model input is trip table, characteristics of the trip and modes available. The output of a mode split model is to produce trip tables for each trip type by each mode. The mode split model output has an impact on air quality, congestion and the connection between land use and transportation. Some of the modal split models used today are based on multi-nominal and nested logit models. This is a sample binomial logit mode split model equation.

69

Auto Utility Equation

$$Us = 1(IVT) + 2.5(OVT) + 0.33(COST) \qquad \ldots\ldots (13\text{-}1)$$

Transit Utility Equation

$$Up = 1(IVT) + 2.5(OVT) + 5(WAIT) + 10(XFER) + 0.33(COST) \quad \ldots. (13\text{-}2)$$

Where,

Us	= Auto Utility
Up	= Transit Utility
IVT	= In Vehicle Time
OVT	= Out of Vehicle Time
COST	= Out of Pocket Cost
WAIT	= Wait Time
XFER	= Number of Transfer

Figure 13-4: Trip Length Frequency Diagram.

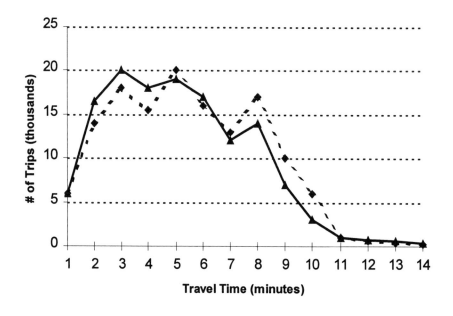

The last step in TDF process is to conduct traffic assignment. As the name implies, both the auto and transit trips estimated are assigned to various routes into the coded highway network. Basically a highway network for modeling consists of all the major freeways, major arterials and secondary arterials drawn using suitable software. This trip assignment output is the final output from the model building process and the results are checked thoroughly using observed data. The ability of the model to replicate observed data becomes the basis for validation for the base year (the year in which the model is built) and the use of model for future year (some chosen year in

70

the future). The input for highway and transit assignment includes a coded highway network and person trip tables produced using earlier trip distribution output.

During the traffic assignment stage, traffic can be assigned by time-of-day like a.m. or p.m. peak period. The output of the traffic assignment is to obtain accurate link volumes and to check the congested speeds on the links. The traffic assignment is driven by volume to delay relationships. As traffic volume increases, travel speeds decrease due to increased congestion. There are different techniques used in conducting traffic assignment. They are all-or-nothing assignment (minimum path), equilibrium assignment, incremental assignment and capacity-restraint assignment. The equilibrium method is becoming the state of the practice. The equilibrium assignment recognizes that transportation link impedance is traffic flow level on links. In this method the model attempts to arrange traffic along the links such that no individual trip maker can reduce path impedance by switching routes.

The final TDF process is model estimation, calibration, validation and application. Model estimation is to determine the model parameters i.e. coefficients which provide realistic model results close to surveyed travel data. Model calibration is to adjust the parameters until the model output replicates observed data by each trip type and each mode. Model validation requires comparing the model output with information other than that used in estimating and calibrating the model. In calibrating the model parameters are adjusted such that the model results fall within acceptable range of error. Model application is to use the validated model to provide output for future year scenario. Also, use the model for conducting sensitivity tests with various transportation policy changes. There are many more steps and procedures than described above in developing transportation demand model.

FIELD WORK AND DATA COLLECTION

In developing a transportation demand forecasting model it is necessary to have knowledge of transportation planning, data survey methods, mathematical background and basic knowledge of traffic flow theory and traffic operations. The most important component of modeling is to provide proper input data to the model and understanding human behavior. As the saying goes "garbage in garbage out" will result if proper input data is not provided during the development of a model. Therefore, data survey is a very important part of model development.

The area to be modeled is divided into zones and data for the zones is obtained from several sources. Majority of the data is obtained from household survey and census tract. A census tract is a small statistical subdivision of county with about 2,500 to 8,000 residents. The census tracts are arranged such that they represent relatively homogeneous areas in population, economic and living characteristics. This data is disseminated and used as input data for traffic analysis zones in developing the model.

The zonal demographics required for modeling are population, household (size), and employment by type, auto ownership, income, travel behavior, regional forecasts, and land use pattern and roadway network data. This type of data is collected for each zone in the area being modeled for current year in which the model is being developed and for the future year in which the model has to forecast traffic.

The different input data required at various stages of modeling process as described earlier is collected by various agencies in an urban area. The most important agencies that collect or gather information are the Metropolitan Transportation Authority (MTA), Metropolitan Planning Organization (MPO), State, Federal and County offices. Some of the traffic related data such as traffic volume; respective transportation and traffic departments in the area collect travel speed and network data. Also, many state and federal agencies are responsible for collecting and developing regional forecasts of urban area population, employment and household. The main responsibility of developing a regional model is with the local MPO. They update the model as input data is received for each zone. In obtaining the data the MPO is supported by several agencies providing land use maps, building permit data, zoning changes, freeway and roadway construction data, vacant land survey, major projects in the area and information on adjacent areas outside the urban area being modeled.

FORMULATING CONCLUSIONS AND RECOMMENDATION

1) Obtain a model output for the current and future year model local MPO and examine the results. Does the output make sense from your current knowledge of the area transportation system and demographic knowledge?
2) Compare the current model year output to the future year output obtained from the MPO? Discuss the main differences in the output results and alternatives recommended in the model report.
3) Fill up your daily trip dairy (activity) for a week in the work sheets. Discuss the travel pattern and the modes used.
4) Research and find out what different types of mode choice models are available for use and compare them.

Lab 13
WORK SHEET
TRAVEL DEMAND FORECASTING: TRIP DAIRY

Name: _____ Activity Day : <u>Monday</u>

SAMPLE TRIP DAIRY

Item	Activity/ Purpose	Trip Origin	Trip Destination	Mode Used	Trip Length (miles)	Departure Time	Arrival Time	Trip Time
1	To School	Home	School	Bike	2.5	7:30 a.m.	7:50 a.m.	20 min.
2	To Restaurant	School	Restaurant	Walk	1	12:05 p.m.	12:20 p.m.	15 min.
3	To Home	School	Home	Bike	2.5	4:10 p.m.	4:25 p.m.	15 min.
4	To Shopping	Home	Book Store	Car	3.5	6:30 p.m.	6:45 p.m.	15 min.
5	To Home	Book Store	Home	Car	3.5	7:10 p.m.	7:28 p.m.	18 min.

Note: Create one sheet per day and summarize the result in mode summary table shown below.

TRIP DAIRY SUMMARY BY MODE

Mode Used	Number of Trips Week Day	Number of Trips Weekend	Total
Car	2	0	2
Bus	0	0	0
Rail	0	0	0
Bicycle	2	0	2
Walk	1	0	1
Other	0	0	0

Note: Data entered refers to sample trip dairy shown.

Lab 14

Transit Demand Modeling

STUDY OBJECTIVES

The main objective of this lab is to introduce students to the development of transit demand modeling using the travel demand forecasting process. Introduce the students to the application of transit demand forecasting in transportation planning.

NEED FOR THE STUDY

In 1999 across the United States transit rider ship is increasing at a rate of 3.85% in the first quarter as compared to the same period in 1998 according to American Public Transit Association (APTA). The overall rate of increase in transit rider ship is 13.6% higher in 1999 than it was in 1996. The main reason for increase is that more and more people are choosing public transit as they find it more attractive alternative to automobiles. This trend is expected to continue and the art of transit modeling is gathering momentum. Previously attention was mostly paid to forecasting auto trips using travel demand forecasting techniques as explained under Lab 13.

The transit models and their output are used by agencies who operate buses and rails, Metropolitan Planning Organizations, County agencies and other public interest groups such as Bus and Rail Riders Union, Schools and Businesses. The need for a transit model is apparent as the demand increases for transit, it is necessary to plan and understand the future demand to provide the best possible service for the commuters using transit. Only the largest metropolitan areas (Chicago, New York, Los Angeles etc.) in the country try to develop transit network and models. Transit models can estimate the demand for transit based on model input and determine the number boarding by trip type (HBW/HBO/NHB), peak and off-peak vehicle requirements, total vehicle hours traveled, total miles traveled, operating costs and transit mode share (compared to autos). Also, the model results provide data on air quality and employment accessibility. Employment accessibility refers to identifying the number of people using transit for work trips in a region.

GENERAL OVERVIEW

Transit demand forecasting is similar to highway demand forecasting, in the former we are predicting the number of transit users and in the later we are predicting the number of auto users. A transit network similar to a highway network represents the existing and future transit transportation system in the area being analyzed.

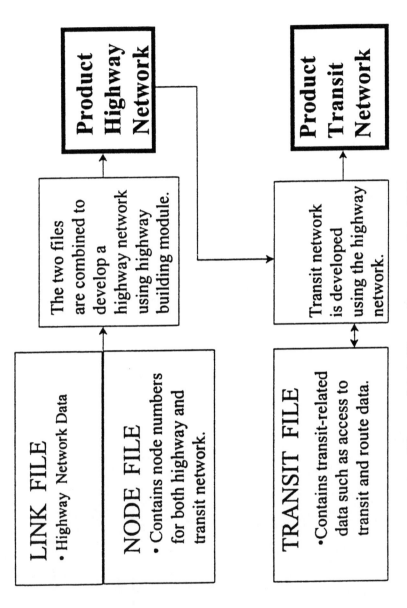

Figure 14-1. Transit network building process.

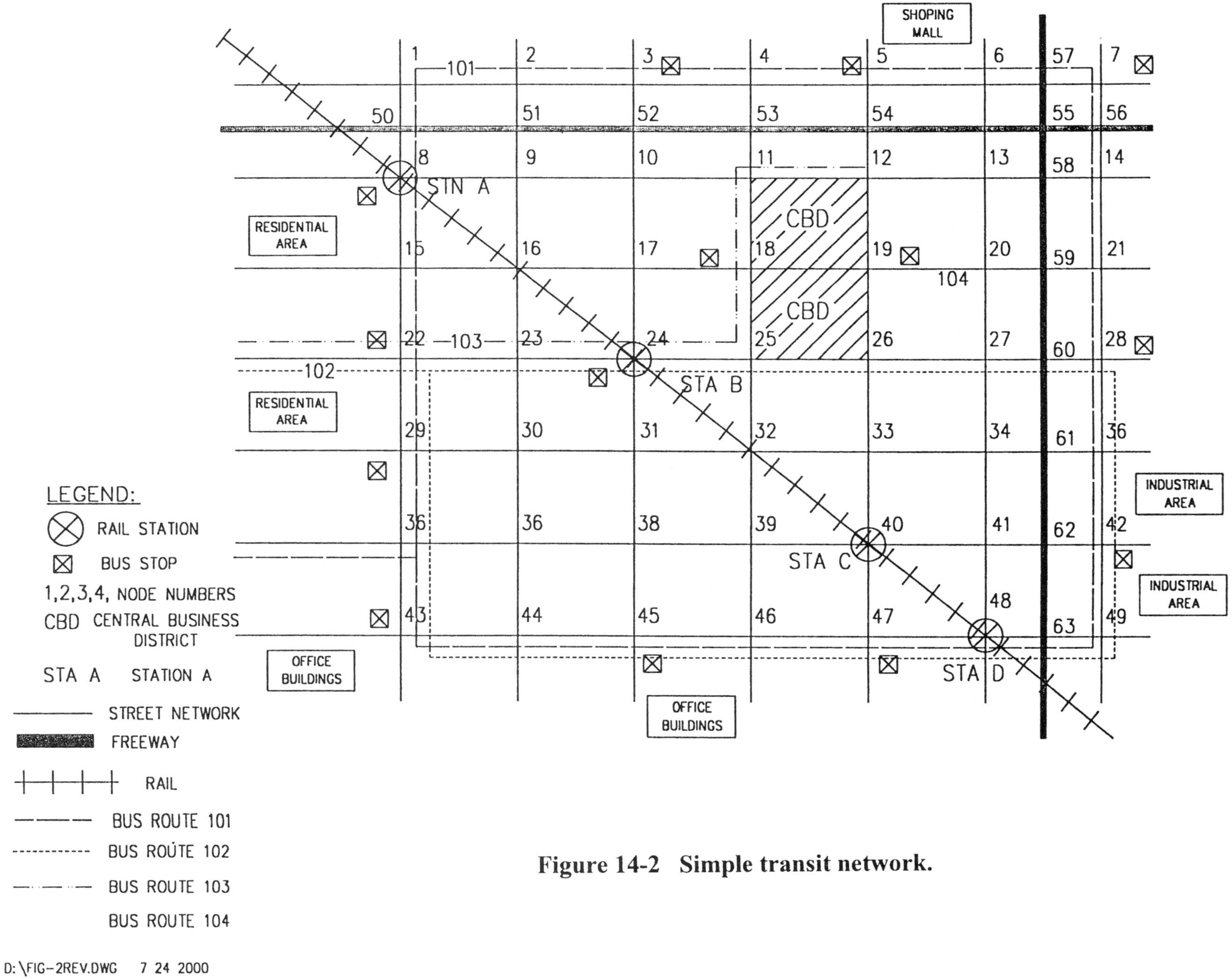

Figure 14-2 Simple transit network.

LEGEND:

⊗ RAIL STATION

⊠ BUS STOP

1,2,3,4, NODE NUMBERS

CBD CENTRAL BUSINESS DISTRICT

STA A STATION A

——— STREET NETWORK

▬▬▬ FREEWAY

┼┼┼┼ RAIL

——— BUS ROUTE 101

–––––––– BUS ROUTE 102

·········· BUS ROUTE 103

— — — BUS ROUTE 104

D:\FIG-2REV.DWG   7 24 2000

76

In developing a transit network both bus and rail (if exists) routes are coded using the highway network. The most common variables of a transit service are the headway, number of stops, route, transfer conditions and fare (cost). In general a transit network is built over an existing highway network. That is the freeway system; arterial system and other road network details created earlier for a highway network are used in creating a transit network. The techniques used to build both the highway and transit network depends on the software being used and input data available at the time of network building. Figure 14-1 shows a typical transit network development in an urban area. The path of bus and rail in the network is traced on the highway network using node numbers as illustrated in Figure 14-2.

STUDY COMPONENTS

An important difference between a highway and transit network is the need to code the access availability to transit. A bus or rail can be taken by walking, bicycle, motorcycle or auto using park and ride facilities. There are specific rules by which access availability to transit should be coded on a transit network. For example for walk access, half a mile is the frequently used limit. Also, walk access links are required at both ends of the transit trip as shown in Figure 14-3. Similarly auto access to transit includes kiss-n-ride and park-n-ride data from stations are input into the model. Auto access is shown at the origin of transit trip as no vehicles are left at the destination end. It is also required to input transfer links and usually they are created by software used for building the transit links. The high cost of data collection and network input limits transit coding work. The next important step in transit network building is to input speeds that represent transit movement in the highway network. For this purpose a look up table, which provides the typical speed for, a combination based on facility type and area type is used. Table 14-1 shows speed input data based on facility type and area type. The next step in transit modeling is path building. That is to find the minimum path between every zone interchange in the network. This is accomplished using mathematical models and appropriate software; the process is similar to steps discussed in Lab 13. Under trip distribution for transit, transit impedance (inter-zonal) are used as input to the gravity model (Lab 13) for each zone interchanges.

Table 14-1 Speed tables for transit network.

Facility/Area Type	CBD	CBD Fringe	Mixed Urban	Non-CBD	Sub./Regional	Rural Area
Freeway	36*	40	42	42	45	60
Principal Arterial	24	27	30	40	40	55
Minor Arterial	22	23	28	36	38	42
Collector Street	22	23	24	27	33	42
Local Street	15	20	20	20	25	30

* All speeds are in miles per hour   Sub.= Suburban

Source: U.S. Department of Transportation, Federal Highway Administration.1998. "Introduction to Urban Travel Demand Forecasting."

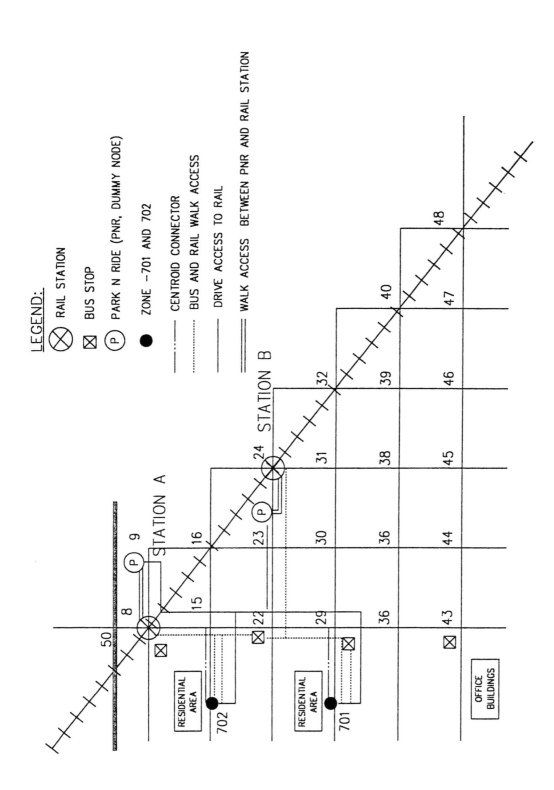

Figure 14-3 Transit network access coding.

A path matrix wills show time, distance and impedance for all zone interchanges. Also, skim tree or travel time matrix is developed which are used during trip distribution, mode share and traffic assignment steps. In summary to develop a transit demand model, an existing highway network is utilized and appropriate transit side input is introduced into the modeling process. The basic four steps (trip generation, trip distribution, mode choice and traffic assignment are followed in estimating transit demand and other valuable transit related output. Table 14-2 shows excerpts from a highway and transit model output.

FIELD WORK AND DATA COLLECTION

Transportation modeling requires large amount of reliable and verifiable input data. For example, in determining the bus route from point A to B along the highway network it is important to obtain the latest route schedule book. The route schedule book should provide the headway, company name, starting time and ending time.

Table 14-2 Sample highway and transit model output.

Description of Model / OUTPUT	Base Year Model	Future Year Model	Alternative One	Alternative Two
	Year 2000	Year 2020	10% More Buses	New Rail Corridor
Daily County Person Trips	27,885,900	31,700,835	31,700,885	31,700,990
Home Based Work (HBW)	5,802,340	6,669,965	6,669,960	6,669,975
Home Based Other (HBO)	15,303,975	16,602,555	16,602,545	16,602,990
Non-Home Based (NHB)	7,777,595	8,450,410	8,420,360	8,428,415
Daily County Transit Trips	814,435	881,900	899,675	885,290
Home Based Work (HBW)	458,300	488,530	504,335	492,055
Home Based Other (HBO)	275,870	304,325	305,315	304,085
Non-Home Based (NHB)	80,256	89,048	90,025	89,154
County Bus Operation Needs	1175	1310	1530	1315
Peak Vehicle Requirement	885	875	888	870
Annual Vehicle Miles	72,755,920	83,090,620	86,702,235	83,863,425
Annual Vehicle Hours	6,778,905	7,560,785	7,814,785	7,622,970
Annual Boarding	337,485,665	324,786,655	343,236,825	324,937,030
Daily Boarding	1,070,525	1,030,240	1,088,760	1,030,710

Also obtain other modeling data such as socioeconomic, demographic and network related data from appropriate agencies. In addition it will be required to update coefficients in model equations used to reflect current environment and make comparisons with similar type of urban areas. Always it is required to seek the latest software and editing tools to save time and budget.

FORMULATING CONCLUSIONS AND RECOMMENDATION

- Consider home as origin and select a destination that is at least 5 to 10 miles by auto, bus and rail (if available) and discuss the differences in travel time (note waiting/idling time), cost (parking, fare and other), routes (walking, park & ride and auto access), safety and comfort
  Suggestion: Each student can select different route in the region for comparison.

- Contact your local Metropolitan Planning Organization and obtain copies of current and future year transit network and model output. Compare and discuss the network and output with your local experience. Discuss if the plan provides benefits to the area?

- Draw a transit network similar to the one shown in Figure 14-2 showing local transit lines (bus and rail, include stops) and access (walk and auto) available for you with home as origin. Similar to Figures 18 and 19 show routes, bike, pedestrian and rail facilities.

- Discuss briefly how effective transit is in your area? How a good transit can alter the future mode share in your area? Include in your discussion the destinations that you think transit should be provided to reduce congestion and improve service in your area.

Lab 14
Worksheet

Route 1

Item	Cost	Trip Origin	Trip Destination	Mode	Trip Length (miles)	Departure Time	Arrival Time	Trip Time
1				Bus				
2				Rail				
3				Subway				

Route 2

Item	Cost	Trip Origin	Trip Destination	Mode	Trip Length (miles)	Departure Time	Arrival Time	Trip Time
1				Bus				
2				Rail				
3				Subway				

Route 3

Item	Cost	Trip Origin	Trip Destination	Mode Used	Trip Length (miles)	Departure Time	Arrival Time	Trip Time
1				Bus				
2				Rail				
3				Subway				

Note: Choose three different destinations at least 5 miles away and compare your transit experience. If you have only one mode just use that mode to complete the tables. Note the number of passengers and any transfers.

Lab 15

Transportation Demand Management

STUDY OBJECTIVE

The main objective of this lab is to provide the student with an understanding of transportation demand management (TDM), various TDM measures and its role in transportation planning.

NEED FOR THE STUDY

In major urban areas such as Los Angeles, Houston, New York, Miami and other large metropolitan city congestion is experienced most of the time, particularly during peak periods. The question to be asked is why does congestion occurs? One of the causes identified is single occupant vehicle (SOV) on our freeways. SOV has impact on both congestion and air quality. TDM is a transportation-planning tool used to reduce the demand on transportation network by reducing the SOV riders by encouraging carpooling, vanpooling and use of transit. The goal is to reduce the peak usage of facilities such as freeways, arterials, parking and other transportation facilities. For example, most universities would like to reduce their parking demand on their campuses. They provide local bus shuttles to student residences. Similarly encouraging commuters to use High Occupancy Vehicle (HOV) lanes.

GENERAL OVERVIEW

Transportation system in any region forms the backbone of that economy and provides a very important service. Transportation Demand Management is a term associated with efficient use of existing transportation network and facilities. The focus should be on the word 'demand,' the goal of TDM is to reduce the demand for existing facilities. Since the 1980's the vehicle miles traveled has been increasing at a much faster rate than population. This is due to high auto ownership, in the 1960's auto ownership was 1.03 per household and in 1980's it is 1.61 per household. Also, over 85% of commuters travel to work by automobiles and of which 83% are SOV. Another phenomenon that is added to the increase in vehicle miles traveled is the development of large suburban areas. The travel between suburban areas is increasing as the job growth in suburban areas is increasing. For the U.S as whole, work trips destined to Central Business District has fallen by 4.5% between 1970 and 1980. The trips to suburban areas have grown by 15%. There is a certain roadway capacity available for all commuters; therefore the TDM strategies developed should be able to increase mobility and air quality in a region. It is not possible for any region to build out of congestion as the financial resources required are limited and cannot risk the deterioration of surrounding environment.

The objectives of TDM are to move as many people from place to place than more vehicles. That is TDM is focused on increasing the person trips than vehicle trips. As seen above, nearly 85% of our commuters use automobiles. How to make a change in behavior? To change the behavior of commuters, TDM relays on incentives and disincentives. One can consider incentives as providing some benefits to the commuters for changing their behavior. Whereas disincentive is to remove some benefits already available to the commuter for not changing their behavior. TDM strategies depending on the goal have to provide both incentive and disincentive in a program. Since the majority of the commuting is for work trips, TDM strategies are developed to be implemented at place of employment. A TDM option will provide many modes for use to increase the mobility, but reduce the number of vehicles on the road. Most TDM programs become a success when the local officials and employers cooperate to implement the alternatives. It has been determined that 30 to 40% reductions in vehicle trips can be achieved by proper implementation of TDM program at a work place. Also, different levels of TDM strategies should be developed depending on different groups requirement, travel pattern and demographics.

TDM solutions are connected to transportation control measures such as transportation system management (TSM, Lab 16). The difference between TDM and TSM is that, TDM tries to change the behavior of an individual commuter, whereas TSM makes the existing transportation system more efficient. TDM is also connected to Transportation Control Measures (TCM's) as defined in federal air quality regulations. A TCM consists of both TDM and TSM strategies. The objective of TCM is to improve mobility and reduce air pollution in a region.

STUDY COMPONETNS

Each alternative and strategy developed under TDM should have a goal and the results of implementation could be evaluated. That is to check whether any trip reduction has been achieved as anticipated. TDM alternatives available to reduce SOV trips are as follows:

1. Compressed workweeks - The employees are allowed to work a full 40-hour per week in less than 5 days, by providing what is known as a 'flex day.' This program is very popular and it is called the 'nine-eighty.' That means to work for 9 days for a total of 80 hours and get an extra day leave, which is generally called the "flex day."
2. Flexible working hours - allow the employees to choose their working hours so as to avoid a.m. and p.m. peak periods of travel.
3. Car pools and vanpool - Provide rideshare information to all employees.
4. Encourage non-motorized commuting such as bicycle and walk by providing incentives.
5. Improve public and private transit network to increase person trips in the area.
6. Telecommuting - allows the employee to work one or more days at home.

7. Construct preferential parking for carpool and vanpool commuters.
8. Provide cash or gift incentive for using carpool, vanpool or transit.
9. Provide regional ride-share hotline - the hotline will provide telephone numbers of local ride-share commuters.
10. Develop bicycle and pedestrian pathways for commuting.
11. Develop road/congestion pricing to reduce SOV.
12. Develop auto-restricted zones to discourage vehicle use and encourage transit use.

In order to implement the above listed TDM alternatives, it is required to develop TDM strategies. The strategies deal with how to attract the commuter and implement the program. TDM strategies may include:

1. Priority treatment for car pool and vanpool riders. Example: Construction of High Occupancy Vehicle (HOV) lanes, preferential parking spaces, preferential entrance and exit points, reduced parking cost and transportation allowances.
2. Public information - that is to develop marketing plans to provide the information in the community. To assure the users that during emergency guaranteed ride home is available. Provide brochures describing the TDM alternatives, local transit schedules, savings due to TDM, and location of park and ride facilities and environmental impact. Under marketing provide the community with overall planning for the area and a vision of the future.
3. Developing parking management programs to discourage SOV.
4. Developing area-wide TDM strategy to discourage SOV by increasing the cost of SOV.
5. Develop innovative TDM alternatives and strategies to reduce vehicle trips.
6. Develop local TDM legislation to implement TDM strategies.

The savings achieved through a TDM program could benefit the society, employer and individual commuter. A TDM program will be successful only when a suitable marketing plan, site amenities and supporting activities are put together and sold to the individual traveler. Figure 15-1 depicts various TDM measures that can be implemented. Table 15-1 shows the economic benefits to the society, employer and individual due to TDM alternative.

Table 15-1 Examples of economics of travel demand management

To Society	To The Employer	To The Individual
Capital and Operating	Direct cost per one-way daily	Cost per one-way daily person
Cost per one-way daily	Vehicle trip removed.	Trip.
Person trip for a 10.5-mile work trip.	Range -$1.95 to $5.62	
	Average $1.33	
Single Occupied Vehicle: $6.75	Net cost per one-way daily	Single Occupied Vehicle: $4.81
Transit: $4.10	Vehicle trip removed.	Transit: $1.82
Carpool (2.5 passengers): $2.70	Range -$3.32 to $4.99	Carpool (2.5 passengers): $1.92
Van Pool (12 passengers): $0.56	Average $0.43	Van Pool (12 passengers): $0.40

Source: U.S. Department of Transportation

Currently, TDM is undergoing some major changes due to new technologies available. That is to provide more incentives towards shared rides, provide universal transit cards (useful in all types of transit within the region), and provide real-time ridesharing data, coordinate with transportation management centers and high occupancy vehicle facility monitoring.

FIELD WORK AND DATA COLLECTION

As described above TDM needs good planning and marketing technique to achieve its goal. In most regions the Metropolitan Planning Organizations and Air Quality Management District in cooperation with other local agencies tries to achieve an 'average vehicle rider ship' (AVR) using TDM strategies. AVR is calculated as the number of persons per vehicle commuting to the site considering all employees using vehicles to travel. That is the ratio to total number of vehicle trips to the site per day by the total number of employees.

Using AVR it is possible to calculate the number of SOV or vehicle trips eliminated and improvements to air quality in the region. Most large employers (over 250 employees) have trained staff who implement TDM strategies and are called " Employee Transportation Coordinators (ETC)."

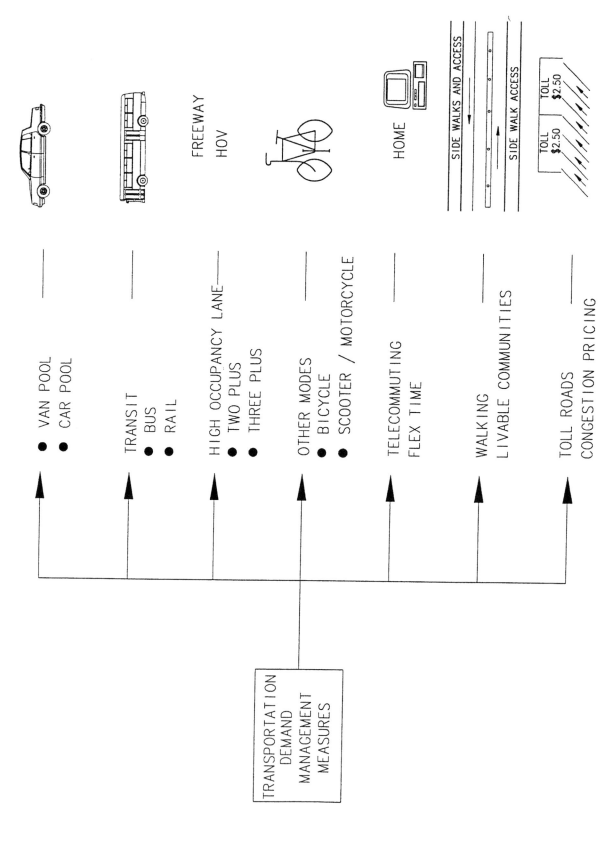

Figure 15-1 Examples of transportation demand management measures.

The fieldwork and data collection is to contact local employee transportation coordinator (ETC's) at various local business (large offices and industry) and request for TDM program adopted (each student can select 3 sites). The data to be collected is the TDM alternatives provided to employees, number of employees, peak and off-peak travel pattern to the site and list of incentives and disincentives adopted. The second fieldwork and data collection is to contact your local Metropolitan Planning Organization and Air Quality Management District and obtain copies of TDM plan for the region. Study the plan and compare to locally adopted strategies by businesses and industries.

FORMULATING CONCLUSIONS AND RECOMMENDATION

1) List various TDM alternatives offered at your school, local business and industries in your area by conducting a mini survey (calling/visiting)?
2) Which TDM alternative is more attractive to the employees from survey and why?
3) What TDM alternative would you prefer and why?
4) Determine how many miles of HOV are available in your county and state? Obtain HOV and SOV (mixed lane traffic) traffic volume data form your local transportation department.
5) What is your opinion on HOV lane use based on traffic volume data collected?
6) List the pros and cons of developing HOV lanes?

Lab 16

Transportation System Management

STUDY OBJECTIVES

The main objective of this lab is to understand, identify and record various transportation system management (TSM) improvements that can be implemented along a survey arterial route to improve mobility and access.

NEED FOR THE STUDY

Transportation System Management deals with ways to manage and increase supply (capacity) of existing transportation system. TSM solutions are very cost effective and can be implemented in a relatively short period of time. TSM improvements play an increasingly vital role in providing short and long-term solutions to congestion relief and improve air quality through increased mobility and reduced delay. Therefore the understanding of TSM is essential for both planners and engineers in improving existing transportation network.

GENERAL OVERVIEW

Transportation System Management is a tool that utilizes a combination of traffic engineering and traffic operation measures to relieve congestion on surface streets for both automobiles and mass transit vehicles. While major capital improvements such as the constructions of new freeways, high occupancy lanes, bridges and roadways are being made in a region, it will be required to provide congestion relief along many existing roadway network. These capital-intensive projects take a long period of time to provide any congestion relief. TSM has gained a lot of popularity because of its relatively low capital cost and the short period of time required for implementation.

The congestion relief is achieved through increased capacity (supply) to move people and goods efficiently, particularly through downtown and densely populated urban and metropolitan areas. TSM improvements do vary greatly depending on local traffic characteristics, design constraints, policies and design standards. TSM solutions cover a wide variety of basic engineering measures ranging from very basic traffic engineering techniques to advanced computerized traffic control and monitoring systems.

The basic traffic engineering solutions under TSM include intersection widening for turning lanes, bottlenecks elimination, marking and channelization, signal

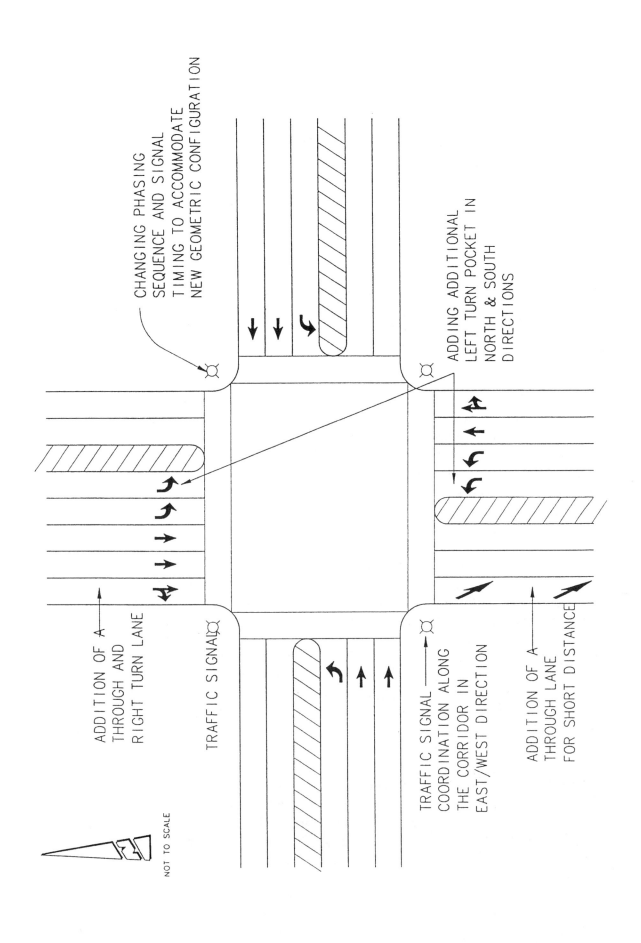

CHANGING PHASING SEQUENCE AND SIGNAL TIMING TO ACCOMMODATE NEW GEOMETRIC CONFIGURATION

ADDING ADDITIONAL LEFT TURN POCKET IN NORTH & SOUTH DIRECTIONS

ADDITION OF A THROUGH AND RIGHT TURN LANE

TRAFFIC SIGNAL

NOT TO SCALE

TRAFFIC SIGNAL COORDINATION ALONG THE CORRIDOR IN EAST/WEST DIRECTION

ADDITION OF A THROUGH LANE FOR SHORT DISTANCE

Figure 16-1 Transportation system management measures at an intersection.

89

coordination, providing priority for transit, controller and cabinet changes, new signal installation, new loop and phasing installation, signing and striping of existing lanes. Figure 16-1 shows typical TSM measures implemented at a busy intersection to improve mobility and access. Also, included are advisory signs such as advanced freeway ramp-exit, directional signing, one-way street, turning movement prohibition and parking restriction.

Under transit preferential treatment measures such as bus turnouts, relocation of bus stops, re-striping of bus lanes, modifications to adjacent arterials, providing exclusive lane for transit, special signal phasing, widening curb radius for easy turning and intersection improvements. Specific traffic operation techniques such as queue jumping, advanced green or green extension, left turn privilege and intersection bus lane is provided.

The freeway TSM improvements include providing High Occupancy Vehicle (HOV) lanes to be used by transit. On arterial streets exclusive bus lanes are provided to decrease delay for passengers. For improving the system it is necessary to integrate local traffic control system with existing traffic signal control system operated by local jurisdiction. It is also important to bring consensus among various local transit operators and affected jurisdictions to make agreements in implementing TSM solutions. The main objective of providing transit preferential treatment is to increase person trips through the corridor.

Advanced TSM solutions include providing computerized traffic control system to increase the performance of signals at intersections and to coordinate the signals along a corridor. Providing multijurisdictional signal coordination will help in increasing the mobility through sub-regions having long corridors with heavy demand. This type of improvements is intended to allow local agencies to expand existing signal control system and surveillance over large area. Computerized traffic control and monitoring system can generally be classified into three areas. They are Arterial System, Closed Loop (Area wide) System and Central System. Also, local jurisdictions can develop their own Traffic Operation Center (TOC). Traffic operation over a multijurisdictional area is conducted by constructing Traffic Management Centers (TMC's).

TSM solution in the future will consists of adaptive traffic control and integrated management system comprising diverse elements and many agencies. In majority of the heavily urbanized congested metropolitan areas advanced traffic control devices are being used on an area or regional basis. The use of Intelligent Transportation System (ITS) will help to manage both recurring and non-recurring incidents on the transportation network. The Intelligent Transportation System (ITS) technology to be adopted under TSM is Advanced Transportation Management System (ATMS), Advanced Traveler Information System (ATIS), and installation of Regional Traffic Management Centers (RTMC) and development of Smart Corridor Projects. Also,

ITS will use Closed Circuit Television, changeable message signs, traffic video imaging, highway advisory radio, computer bulletin, information kiosks and advanced communication (two-way) system.

STUDY COMPONENTS

In conducting a TSM improvement it is necessary to study the existing conditions along a selected route and make recommendations to improve mobility. The TSM improvements can be classified into two categories as follows:
1. Basic Traffic Engineering Solutions (BTES) and
2. Advanced Traffic Engineering Solutions (ATES).

BTES include conventional low cost improvements such as:

1. Intersection widening for turning lanes (through/left/right)
2. Identification and elimination of bottle neck roadway segments or intersections
3. Re-striping, channelization and widening of short radius curbs
4. Signal coordination and new signal installation
5. Signal operation modification/upgrade control equipment/replace control equipment
6. New signal loop installation
7. Capacity enhancement through removal or reallocating parking and/or driveways
8. Providing advance signs based on location needs
9. Relocation of bus stops (to the far side of intersection) and re-striping of bus lanes
10. Operational changes to facilitate passage of transit at intersections such as queue jumping, green extension, advanced green, left turn privilege etc.

ATES include projects that are capital intensive such as:

1. Signal priority system (bus speed improvement) along a route.
2. Use of vehicle identification technology and GPS
3. Computerized traffic signal control and monitoring system
4. Traffic operation centers (TOC's) for local application.
5. Traffic Management Centers (TMC's) for region wide application.
6. Development of smart corridors on selected routes
7. Use of ITS technology such as ATMS and ATIS.
8. Installation of Closed Circuit Television (CCTV's), Changeable Message Signs (CMS), Highway Advisory Radio (HAR) and Advanced Vehicle Locating (AVL) system.
9. Installation of information kiosks and high technology two-way communications.
10. Telephone call-in system and computer bulletin boards.

As a rule of thumb projects requiring a total cost between $100,000 to $500,000 can be classified as low cost projects and all other projects over $500,000 can be classified as capital-intensive projects. For some smaller local jurisdictions even a $100,000 could be a high cost project, like installing a new signal at an intersection.

In identifying and classifying the various solutions recommended under TSM, it is essential to consider the cost of each improvement recommended. Majority of the time ATES are carried out in phases over several years based on funds availability. Whereas some of the BTES solutions such as re-striping or intersection widening can be carried out immediately to relieve traffic congestion and increase mobility.

FIELD WORK AND DATA COLLECTION

TSM improvements are recommended for arterials experiencing delays and congestion. Select an arterial that has at least six to eight intersections (more the better). Drive along the route to see whether the signals are coordinated. Even if the signals are coordinated there will be several other TSM improvements that can be recommended to reduce delay along the test route. Observe the delays at each intersection and the type of delays (see labs 2 and 5). Also, note the dimensions of the street, peak and off-peak traffic volume (obtain from local jurisdiction), posted speed, pedestrian crossing, lane configuration, and location of left and right turns pockets and driveway locations. Use the worksheets to record your field data.

After completing the field survey, analyze the data to recommend TSM improvements that will increase mobility and reduce delay. Select the best possible TSM solutions and record them on work sheet.

FORMULATING CONCLUSIONS AND RECOMMENDATIONS

1) What are the existing geometric and intersection characteristics along the study route?
2) What is the estimated delay (using data based on labs 2 and 5) and what TSM solutions are available for reducing delay along the survey route?
3) Classify your improvements into BTES and ATES?
4) What are some of the BTES that you would recommend for immediate implementation?
5) Estimate the total cost of BTES and ATES for the test route?

Lab 16
Work Sheet
Transportation System Management

DATE_____ DAY_____ WEATHER_____
PAGE_____

STUDENT NAME (S) _____

ARTERIAL NAME: _____ ORIENTATION_____

NUMBER OF LANES_____ POSTED SPEED_____

NUMBER OF INTERSECTIONS ON TEST ROUTE _____

STARTING TIME_____ ENDING TIME_____

Field Survey Data: Intersection _____

Intersection	Total Peak Hr. Vol.	Total Off. Pk. Hr. Vol.	# of Through Lanes	# of Left Turn Lanes	# of Right Turn Lanes	Comments
1 St. and A St.	1249	348	3	1	None	Heavy right turns
						observed during peak
						period.
						Left turn pocket
						blocks left through
						lanes westbound.

Improvement Recommended: 1 St. and A St.

#	Intersection Name	Improvement Recommended	Estimated Cost	Classification
1	1 St. and A St.	Provide exclusive right turn lane WB, lane width available only striping of right lane is required.	$1,800	BTES
2				
3				
	Total Cost			

Improvement Recommended: Tustin Avenue

#	Arterial Segment between	Improvement Recommended	Estimated Cost	Classification
1	Serrano Av. To Canyon Rim (2.4 miles and 5 intersections)	Needs signal synchronization and a new signal at Sunset Ridge Road	$130,000	BTES

93

Lab 17

Traffic Impact Studies

STUDY OBJECTIVES

The main objective of this lab is to understand the fundamentals of traffic impact studies (TIS) conducted to determine the impact of land use on surrounding roadway network. To study the use of TIS in environmental impact report (EIR), developer impact fee and other transportation planning projects.

NEED FOR THE STUDY

Whenever there is a new land use (houses, school, office building and strip mall) or replacement of an existing land use (includes expansion) the project site will generate traffic onto the surrounding roadway network. To understand the impact of this new traffic, it is required to have traffic impact studies. TIS studies are conducted to document the impact of new traffic on existing and future traffic roadway network within the influential area from the project site. Such studies also determine the access (ingress and egress) to the site and number of parking spaces required. The objective of a traffic impact study is to reduce the traffic impact due to the project to an acceptable level of service (LOS). It is achieved by recommending suitable solutions such as adding turning lanes, through lanes, additional parking and signals.

In case of projects such as large shopping malls, industrial projects, master planned residential areas etc. the local planning agencies may require a full environmental impact report which will have a traffic impact section. Based on the determination of traffic impact due to a project, the local jurisdiction may impose a developer fee to implement recommended solutions to maintain acceptable LOS.

GENERAL OVERVIEW

A traffic impact study will address various issues related to transportation and traffic operations due to a new or replacement development on a project site. The major traffic related issues are 1. To determine "How much traffic does the project generate? and; 2. What impact does the project generated traffic have on surrounding street network?" Also the study should address short and long term impact of the project with regard to access, off-site impacts, on-site circulation, and non-site traffic and delivery routes. Overall TIS assists both developer and local agency to make suitable land use and transportation plan.

When is a traffic impact study required for a project? In general TIS will be required whenever a project generates substantial peak period (a.m. or p.m.) and total daily traffic. The threshold for conducting TIS depends on each local jurisdiction guidelines. Basically trip generation by the project determines the need for TIS. Other reasons for conducting TIS may be due to the potential impact on existing neighborhoods, proximity of driveways, anticipated cumulative traffic impact (combining traffic from other projects in the area) on the area and certain specific project type (landfill, sports arena, theme park and hospital).Once the need for traffic impact study is determined for a project. The next step is to determine the study area limits. That is to determine the area of impact of the project traffic on surrounding roadway network. In general the local agency will determine the area of influence that includes all intersections and roadway links that the project traffic may use for inbound and outbound from the project site.

The next step by the local jurisdiction is to identify what extent the study should be conducted. This step provides detailed description of the TIS contents required by the local jurisdiction. It includes details regarding traffic counts; other project traffic in the area, intersection geometric requirements and technical analysis required. At the time of preliminary meetings the designated developer traffic person (Project Manager) and local jurisdiction staff will briefly discuss issues such as trip generation, trip distribution and traffic assignment due to the project. Also, if the project has various phases, the phases will be discussed in detail.

The developer obtains the details of the project requirements. The next step is to contact other agencies that may have influence on the project. As mentioned earlier the extent of TIS depend on the location, size, type of development and existing traffic conditions surrounding the project. Smaller projects may need minimal study, whereas large projects generating large daily trips will require extensive studies.

The final element of a TIS project coordinator is to discuss the project contents and the format in which the report is to be produced for distribution. If the project requires an Environmental Impact Report (EIR), the time required to prepare such a report will be long and many agencies need to be contacted before final report is prepared and accepted. During the preparation of the TIS it is essential to learn about any local issues that influences the project study.

The following provides the basic contents of TIS study report:

1. Detailed project description with location map and any other adjacent project(s) location (Figure 17-1).
2. Existing traffic volume at study intersections and links. Tables showing project generated traffic during peak and off-peak periods. For trip generation for projects the "Trip Generation Hand Book" published by Institute of Transportation Engineers (ITE) is used.

95

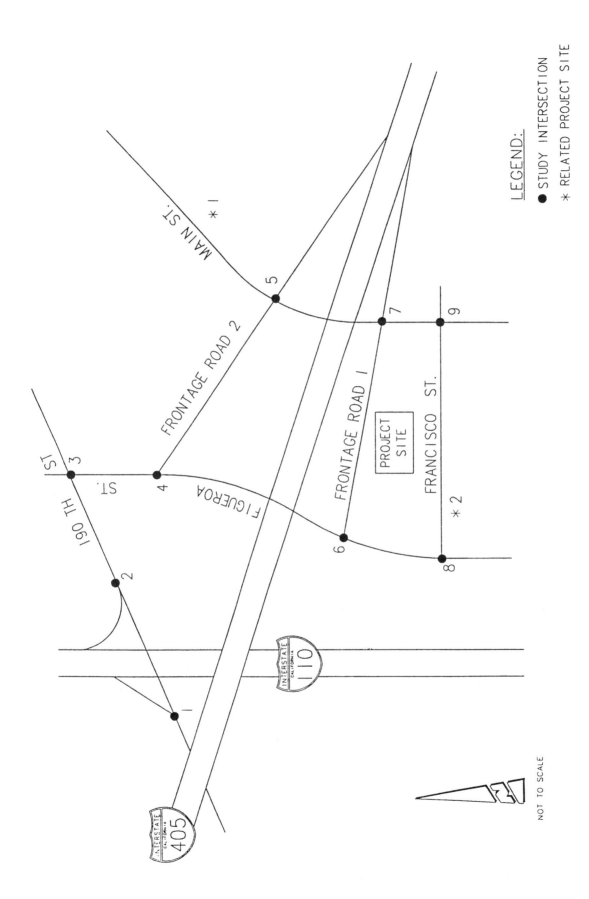

Figure 17-1 Project location and study intersections.

LEGEND:
● STUDY INTERSECTION
* RELATED PROJECT SITE

NOT TO SCALE

96

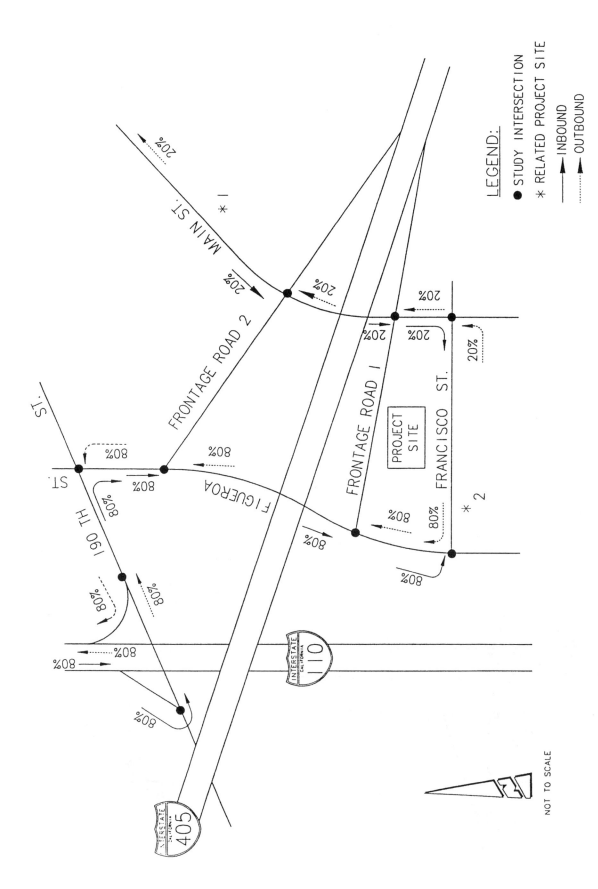

Figure 17-2 Project trip distribution inbound and outbound traffic.

NOT TO SCALE

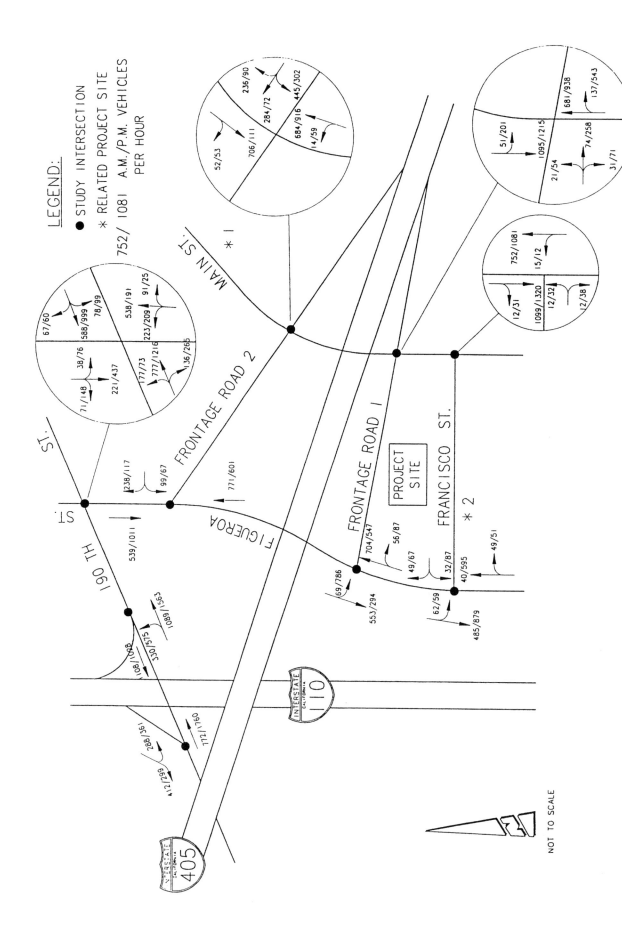

Figure 17-3 Cumulative a.m. and p.m. peak periods traffic volumes.

3. A map showing the trip distribution (Figure 17-2) of the project traffic, cumulative traffic from the site (Figure 17-3, includes non-site and other projects traffic in the area).
4. Analysis showing the level of service of study intersections and links due to existing, existing plus growth factor (if given for the area), existing plus growth factor plus project traffic and finally the cumulative traffic that includes existing plus growth plus project plus other project(s) traffic in the study area. The analysis will include any mitigation recommended to meet the local jurisdiction requirement for level of service.
5. Maps showing cumulative traffic passing through the study intersections and links (ADT).
6. The report should describe the above steps in detail with appropriate references. Also, the study will provide details of driveway, signal warrants (if required), parking, on-site and off-site improvements, implementation timing, mitigation cost and other related issues of the project.

Who conducts such studies? Private transportation and traffic consultants on behalf of the developer generally conduct it. The study will be conducted based on a set of guidelines (local, state and federal) provided by the agency requesting the study. The fee for conducting TIS varies and depends on the requirements of a study.

FIELD WORK AND DATA COLLECTION

In conducting a TIS field visit and traffic data collection are major tasks. The fieldwork relates to visiting the project site and documenting the details of influential area recommended by the local jurisdiction for the study. Details such as the number of study intersections, roadway links, driveway locations per plan (ingress and egress from site), traffic counts (peak and off-peak), location of other projects (for cumulative analysis), number of signalized and unsignalized intersections and parking (on-site and off-site). Other types of traffic related data that may be required based on the study requirements and project type. Some other traffic related data that may be included in the notes during site survey is to make note of posted speed, bus stops, sight distance (vertical and horizontal curves with problems) and rail crossings.

For most TIS machine traffic counts are taken on roadway segments to obtain average daily traffic. Intersection turning movement counts are taken during a.m. and p.m. peak periods to conduct capacity and LOS analysis. In collecting traffic data and preparing the report the preparer should ensure that all site data, traffic count data and other related work are per current practice of preparing a TIS. Attention should be paid in estimating both the project trip generation (using the ITE Trip Generation Handbook) and non-site traffic (passing trips). Because trip generation is the most critical data provided in a TIS.

The next step is trip distribution of the project traffic. The first step here is to conduct a field review to observe the pattern of flow in and around the project site. Trip distribution is based on local knowledge, project trip attraction characteristics, surrounding land use and population distribution and if required using transportation model for large projects. The final step is traffic assignment, which is based on trip generation and trip distribution of the project. Trip assignment allocates number of trips made in and out of the project site on the identified roadway network. The final stage of TIS is to conduct capacity analysis based on the requirements of the local jurisdictions. The report should use proper technical analysis per local jurisdiction requirement. The level of service at intersections is conducted using available software and current practice. The threshold level of service is defined by the local jurisdiction. A project is usually said to have impacted an intersection or roadway when the addition of project traffic to existing and growth factor traffic exceeds the threshold level of service adopted by the local jurisdiction. Mitigation is to be recommended to bring the level of service equal to or below the threshold level of service.

FORMULATING CONCLUSIONS AND RECOMMENDATIONS

1) From your local city Traffic Engineering Department or a traffic consultant firm request for a copy of a traffic impact study report. Study the report and write a summary on the report describing problems and solutions recommended.

2) Take manual turning movement counts at a busy intersection close to your school for fifteen minutes during a.m. and p.m. peak periods and discuss the flow pattern based on trip generators (facilities) located along the main corridors.

3) Contact at least three jurisdictions (cities) in your region and obtain copies of TIS guidelines and compare the requirements.

4) Determine the ingress (entering) and egress (exiting) pattern of traffic from local fast-food restaurants driveway using worksheet shown?
Suggestion: Conduct a survey for thirty minutes at each driveway and record the pattern of flow.

5) Based on your survey in question 4 do you recommend any improvements to the driveways and street network? Draw simple sketches to show the improvements recommended.

6) Check with your local jurisdiction if there is a developer fee imposed for developments in your area and list the impact fee for various types of developments (residential, office, shopping malls etc.).

Lab 17
Work Sheet
Traffic Impact Studies

Trip Generation: Fast Food Restaurant

Time Period	McDonald's		Burger King		Carl's Junior	
	Number of Vehicles		Number of Vehicles		Number of Vehicles	
	Entering	Exiting	Entering	Exiting	Entering	Exiting
A.m. (7:00 - 8:00)						
m.d. (12:00 - 1:00)						
P.m. (5:30 to 6:30)						
Area	1,300 Square Feet		800 Square Feet		1,000 Square Feet	
Characteristics						
Drive Through	Yes		No		Yes	
Located in strip mall	Yes		Yes		No	
Detached	No		No		Yes	
Along Primary Arterial	Yes		Secondary Arterial		Yes	
Close to Freeway Ramps	No		Yes		Yes	
Number of Driveways	2		1		3	
Traffic Volume	Adjoining Street					

Note: Same format can be used for other facilities to determine trip generation data.

Conduct survey during both peak and off-peak.

By selecting different facilities you can compare the trip attracting characteristics.

Lab 18

Transportation and Air Quality Management

STUDY OBJECTIVES

The main objective of this lab is to understand, identify and record basic impacts of transportation system on air quality. To learn why air quality management is required to reduce the impacts of transportation on our environment.

NEED FOR THE STUDY

Around the world transportation forms the backbone of any economy. Particularly so in the United States, we have the largest and the most efficient transportation system in the world. In the parking industry they have a saying, which well fits the attitude in using cars in the US, the saying goes like this "A pedestrian in America is one who got parking." When a vehicle is idling or running it is emitting combustion materials into the atmosphere causing pollution. Air pollution is a very vast and complicated subject.

How is human health affected by air pollution? Majority of the air pollutants is inhaled during breathing. Once the air pollutant has entered the body and reached lungs, later it enters brain and other parts of the body. For example, carbon monoxide emitted by autos is absorbed through blood. Carbon monoxide has high affinity for hemoglobin (iron-based organic compound), but hemoglobin is required by the body to transport oxygen in the blood. Thus oxygen is eliminated from the cells and thus leading to health hazard. The absorption of air contaminants depends heavily on the size of particles, as smaller size particles are more easily absorbed and have greater effect on health.

In general lungs, central nervous system, liver, kidneys, blood and reproductive system are affected by both criteria and non-criteria pollutants. The health effects could be allergic reaction, asphyxiates, irritants, necrotic stimulation and systemic poisons. There will be different degree of reaction to the pollutants by each individual based on age, sex and amount of exposure. For example, children are more sensitive to pollutants, because their body system has not fully developed. Similarly, the elderly are more affected as the body system has deteriorated over the years.

GENERAL OVERVIEW

Congress passed the original Clean Air Act (CAA) in 1970 and it was only 65 pages. The Clean Air Act Amendments (CAAA) of 1990 is nearly 800 pages of which 40 pages of it contain transportation related provisions. The substantial increase in documentation between 1970 and 1990 can be inferred as an indicator of the advancement and understanding air pollution as a serious threat to public health. The CAAA was necessary because it was clearly established that the anthropogenic (human made) emissions needed to be reduced on which we have control. That is developing air quality management strategies.

The primary sources of pollution are the Geogenic, Biogenic and Anthropogenic. Geogenic and Biogenic sources refer to natural causes of air pollution. For example, Geogenic refers to emissions from volcanoes and Biogenic refers to emissions from plants. Anthropogenic refers to man-made air pollution due to cars, industries and other activities. Next the sources are divided into mobile and point sources. Mobile sources are those due to vehicular movement such as use of vehicles for travel. Point source refers to a place or area where air pollution is occurring due to an industry (fumes) or construction (dust). Over the years with dedicated research we have been able to differentiate the pollutants that are the cause of immediate health hazard for all humans. The pollutants are further classified into two categories as criteria pollutants and non-criteria pollutants. There are six criteria pollutants, they are

1. Sulfur dioxide
2. Nitrogen dioxide
3. Carbon Monoxide
4. Lead
5. Particulate Matter
6. Ozone

The remaining pollutants are classified as non-criteria pollutants. Of the six criteria pollutants the most important and widely studied pollutant is Ozone. The formation of criteria pollutant ozone has been attributed to the emissions of hydrocarbons and nitrogen oxides from automobiles and trucks using the roadway network. The other mobile sources are ships, rail, airplanes and buses. Ozone is an invisible odorless gas that is formed through complex chemical reactions and it is the major component of what is called as SMOG. The term SMOG was coined using the two words 'smoke' and 'fog' in Los Angeles, Southern California in the early 1960's. The formation of ozone in the atmosphere is due to the presence of nitrogen oxides (NOX), hydrocarbons (HC) and other reactive organic compounds (ROC's) emitted by automobiles and other mobile sources. The ozone concentration tends to be higher when solar radiation is most intense i.e., on hot days. It has also been observed that the formation of ozone in the atmosphere follows the morning (a.m.) and late (p.m.) peak period travel pattern. The a.m. and p.m. normal travel pattern of automobiles is shown in Figure 8 in Lab 4. Figure 18-1 shows that the criteria pollutant Carbon Monoxide has maximum output during the a.m. and p.m. peak periods.

What does smog do? When smog is inhaled it damages the cells in the lungs, airways, making air passages inflamed (swollen). In addition it can reduce the respiratory systems ability to fight infection and remove foreign particles from lungs and air passages. Apart from causing harmful effects on humans, smog can also affect vegetation, crop degradation and leaf damage. For example, in the Southern California the Southern California Air Quality Management District (SCAQMD) has estimated that it costs $1.60 per day due to health and damage to vegetation, paint, rubber and other materials exposed to atmosphere. In terms of quantification it is estimated that two-thirds of all air pollution is due to automobiles in most regions as seen in Figure 18-1. The impact of vehicle idling, acceleration, deceleration and cruising on air pollution are shown in Figure 18-2.

Table 18-1 U.S. emissions by source classification.

Source	CO	PM10	NO	SO2	VOC	LEAD
Transportation	43.5	1.51	7.3	1	5.1	1.62
Combustion	4.7	1.1	10.6	6.5	0.7	0.45
Industrial	4.7	1.84	0.6	3.2	7.8	2.21
Solid Waste	2.1	0.26	0.1	0	0.7	0.69
Miscellaneous	7.2	0.73	0.2	0	2.6	0
Totals:	62.2	5.45	18.8	20.7	16.9	4.97

*In millions of tons per year, except Lead (thousands). Source: Natural Air Quality and Emissions Trends Report, 1991.

Table 18-2: Engine emission factors.

Condition	CO (gr./sec.)	NOX (gr./sec.)	ROG (gr./sec.)	Time (sec.)
Idle	0.00191	0.00124	0.0012	
Cruise	0.00488	0.00945	0.00334	
Acceleration	0.06781	0.02178	0.01155	11.62
Deceleration	0.00177	0.00256	0.00119	12.25

Gr. = Grams   Sec. = Seconds

Source: Texas Transportation Institute. "An Emission Model for Arterial Streets." 1992.

The primary objective of the 1990 CAAA was to overhaul of the planning provisions for those areas in the country not meeting National Ambient Air Quality Standards (NAAQS) shown in Figure 18-3. The law identifies specific steps for emission reduction, requires local governments to show reasonable progress and an attainment demonstration, and incorporates more stringent requirements including penalties for not meeting interim goals and milestones.

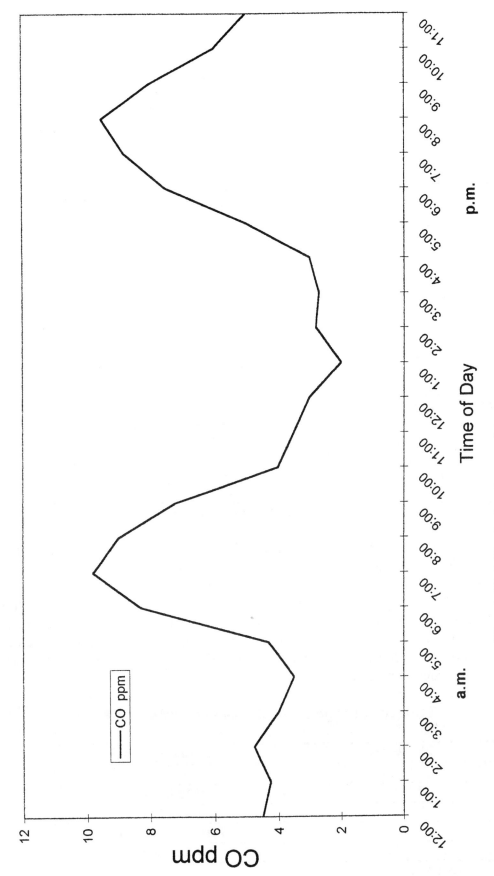

Figure 18-1 Typical urban carbon monoxide levels.

Source:Principles of Air Quality Mangement, Lewis Publishers.

The CAAA rely on the development of transportation Control measures (TCM's). Examples of TCM's are construction of High Occupancy Vehicle (HOV) lanes, development of transportation demand management (TDM) programs, and traffic flow improvements such as transportation system management (TSM), expand transit routes etc. The objective of TCM's is to improve; 1. The operational efficiency of the existing transportation system 2. Relieve congestion and 3. Curb the growth of vehicles miles traveled. The air quality management districts are required to prepare 'Air Quality Management Plan' to clean up the environment per both Federal and State laws as applicable.

Table 18-3: National Ambient Air Quality Standards (Criteria Pollutants).

Containment	Concentration *	Duration (hours)
Ozone	0.12	1
Carbon Monoxide	35	1
Nitrogen Dioxide	0.05	annual
Sulfur Dioxide	0.14	24
	0.03	annual
Particulate Matter - PM10	150 mg/m3	24
	50 mg/m3	annual
	1.5 mg/m3	3 months

*In parts per million (volume), unless otherwise indicated.

From the point of public understanding of air pollution Environmental Protection Agency (EPA) has determined that a normalized reporting system that is easily comprehended by the public. The pollutant standard index (PSI) integrates information for the different criteria pollutants, except lead (as it does not have a short-term NAAQS) across an entire monitoring area in to one normalized figure. The PSI levels used are shown in Figure 18-4. This index in ranging in numerical values from 0 to 500 is intended to represent the best and worst air quality in the monitoring area respectively.

Table 18-4: Pollutant standard index levels.

PSI Range	Air Quality Level
0 - 50	Good
51 -100	Moderate
101- 200	Unhealthful *
201 - 300	Very Unhealthful*
>300	Hazardous*

* EPA Designation

The Intelligent Transportation System (ITS) technology supported by air quality management are Advanced Transportation Management System (ATMS), Advanced Traveler Information System (ATIS), installation of Regional Traffic Management Centers (TMC) and development of Smart Corridor Projects. The use of Intelligent Transportation System (ITS) will help to manage both recurring and non-recurring incidents on the transportation network and avoid delays and idling of vehicles thus assist in reducing air pollution.

STUDY COMPONENTS

How is air pollution transmitted? The transport of air contaminants depends on the geographic nature of the area and wind factor. The geographic nature describes whether the area is plain, mountainous, rolling or a combination of the three types of terrain. The wind factor describes how strong the wind blows in the area, where it is generated? Because the temperature of wind could vary depending on where it is generated, i.e., over sea, desert or other varying geographic areas. Also, the climate of the area plays an important role in the dispersion of air contaminants through a region. On a global scale, oceans, mountains and continents creating high and low pressures affect the movement of wind or jet streams. Also, it will vary from latitude to latitude across the globe and affected further by earth's rotation.

The phenomenon of inversions is an important meteorological effect that affects air pollution in a region. At higher altitudes the temperature of air is found to be higher than it is on the surface. There is a certain height at which the temperature of air stays higher than at surface and this layer of air is called inversion layer. What is the impact of an inversion layer? The impact is that it does not allow the mixing of air moving upwards, i.e., it forms a barrier at higher altitude for polluted air from urban and sub-urban areas to mix and disperse. There are various types of inversion. In Los Angeles basin the marine layer that forms over the Southern California coast gets transformed into smog due to the presence of inversion layer, large-scale auto emissions and surrounding mountains, particularly on hot days.

 Government regulation becomes an important element in controlling air pollutants. The requirement for states to develop 'state implementation plan (SIP)" is the basis for air quality management. The SIP illustrates the plan and records ongoing improvements in air quality standards across the state. Majority of the SIP is a compilation of all regional air quality management districts plans. If an area is found to exceed ambient air pollution limitations, a federal emergency can be declared region wide to shut down industries, schools and commercial establishments. There are several laws (titles) and regulations, related to various criteria and non-criteria pollutants at both Federal and State levels and includes outdoor and indoor air quality.

FIELD WORK AND DATA COLLECTION

Fieldwork and data collection in air quality management involves the review of materials blown in to the atmosphere by autos, industries, commercial establishments, construction and daily activities. It is possible to estimate the tonnage of the various materials such as mercury, cadmium, arsenic, lead, zinc, sulfur dioxide, carbon monoxide, chromium, chlorine, barium, nitrous oxide etc. depending on the source (mobile or stationary). Usually the emissions are estimated and tabulated as tons/year or tons/day with the source identified.

In the filed it is essential to measure the air pollution due to transportation sector due to idling, cruising, accelerating and decelerating. Using clean fuels such as compressed natural gas, electricity and fuel cells can reduce emissions such as NOX and particulate from mobile sources. Also, air quality could be improved through transportation control measures such as providing HOV lanes and signal synchronization. By conducting a 'before and after' study it is possible to estimate the reduction in air pollution due to an employed strategy.

The cost estimation to remove certain pollutant will be based on the total budget required for air quality management strategy, regulations, planning and permitting requirements. As a result of this, majority of the air quality strategy implementation will be based on "fee" to be paid by the industry or client. To reduce the cost of control, it may be required to develop certain standards at the manufacturing level to meet stringent air emission standards throughout the life of the equipment. Also, equipment's can have automatic indication to warn when the equipment has exceeded current limits for particular air pollutant. Thus requiring replacement.

FORMULATING CONCLUSIONS AND RECOMMENDATIONS

1) Contact your local Air Quality Management District and obtain a recent year Air Quality Management Plan (AQMP)? Discuss in the air quality plan of action (remedial solutions) to reduce air pollution in your region.
2) List the traffic control measures in the AQMP for your region and how does it change the emission inventory from current and future according to the plan?
3) Compare the federal and your state acceptable limitations for criteria pollutants?
4) List the total cost of implementing the AQMP solutions and sources for funding.
5) Choose a major or a secondary arterial in your area and select a segment (at least 6 intersections) of an arterial with signalized intersections. Using emission factors determine the total pollution due to idling, cruising, accelerating and decelerating (Figure 18-2) in traveling this section.

Obtain the average daily traffic on this section and determine total pollution per day.

6) List what action would you recommend educating the public about reducing air pollution?

7) What personal actions are you willing to take to reduce air pollution, list the actions?

Lab 19

Transportation Funding

STUDY OBJECTIVES

The main objective of this lab is to provide general information on various types of transportation funding and programs available for various transportation projects. Also, to provide an insight into how transportation projects are ranked and funded.

NEED FOR THE STUDY

A common thread that connects all transportation projects and activities is transportation funding. Where does the project funding come from? How are funds raised for these projects? For future transportation professionals it is required to understand how transportation projects are funded and implemented on the field. Majority of transportation projects such as construction of freeways, arterial streets, bridges and other infrastructure require millions of dollars. Without funding no transportation project could get off the ground. There are several funding sources from federal, state and local agencies. The funding is allocated through various programs for variety of transportation projects. Also, new funds are raised based on a regions need by local governments. Without funding there is no project, so understanding how funds are obtainable is essential for a successful project.

GENERAL OVERVIEW

Selecting transportation projects for funding is an important task of any transportation agency such as Metropolitan Planning Organizations and Metropolitan Transportation Authorities with responsibility to receive and distribute these funds for projects. In general funds come from federal, state and local agencies. Figure 19-1 shows the general transportation fund flow from federal to local level. Under federal government there are two main agencies, which are responsible for allocating funding for transportation projects. They are 1. Federal Highway Administration (FHWA) and 2. Federal Transit Authority (FTA). The FHWA and FTA have several internal departments for the various transportation projects, which have the responsibility to evaluate projects for funding. From federal the funding comes to designated state's main transportation agency, generally the state transportation department. The state administers all the funds received from FHWA. For FTA funds local agencies apply directly to FTA. The state agency will distribute the funding to local agencies through counties and cities through a competitive process. Majority of the large

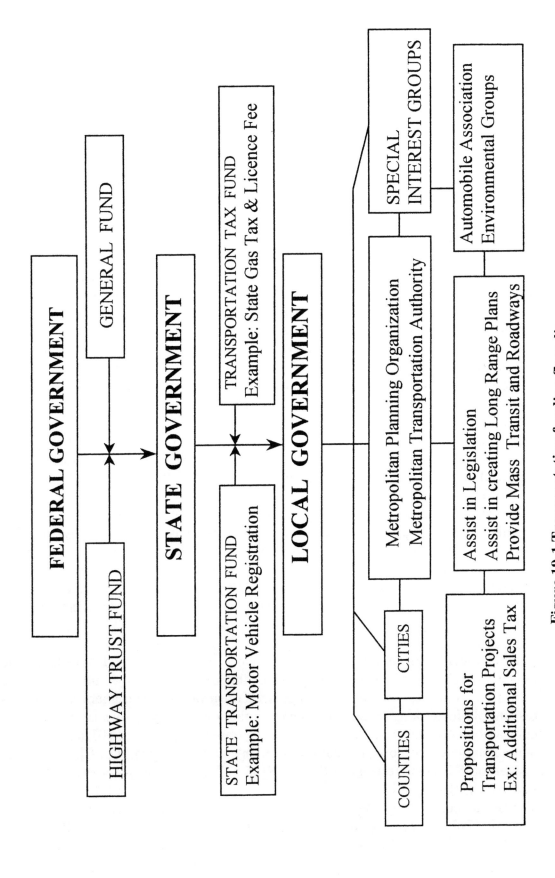

Figure 19-1 Transportation funding flow diagram.

counties have their own metropolitan planning and transportation agencies, which are designated to plan the transportation system and distribute the funding to local cities. Also, some funding is allocated to special interest groups dealing with specific transportation projects that are beneficial to the public.

STUDY COMPONENTS

The following section briefly describes the various funding and plans available. The most important are the federal, state and local funding.

Federal Funding

Federal government collects funding amount through federal gas tax, federal diesel tax, tire and truck sales. Also, it collects taxes on several user taxes and special fuel taxes (gasohol, special fuels, ethanol and methanol). Federal government has two accounts 1. Highway Account and 2. Mass Transit Account. In order to distribute the collected funds, the federal government has passed several legislation's such as the Transportation Equity Act (TEA-21, 1998), Air Quality Act (1990), American Disability Act (1990), National Environmental Policy Act (1969), Urban Mass Transportation Act (1964) and Federal Aid Highway Act (1956)

In general FHWA develops guidelines for many of the transportation programs. The projects are sent by the states under State Transportation Improvement Plan (STIP) for selection at the federal level. The selection is based on specific criteria developed by FHWA. Once a project has been ranked and selected the project funding is allocated. The funding is transferred to the state account for distribution. FTA has slightly different approach; the local agencies can apply directly for use of FTA funds. Also federal government develops several funding plans which act as guidelines. Some of the plans are Transportation Improvement Program plan (TIP), Federal Transportation Improvement Plan (FTIP), Federal Statewide Transportation Improvement Program (FSTIP) and Regional Transportation Improvement Program (RTIP).

The Intermodal Surface Transportation Efficiency Act of 1991 (ISTEA 1991, Public Law 102-240) is a federal law that provides funding for highways, transit systems, ferries, roads, bicycle facilities and other types of surface transportation. ISTEA was responsible for making sweeping changes to these programs in over 30 years. The emphasis was to look at more choices, making transportation more efficient, improve non-automobile transportation and invest in enhancing our quality of life. The total budget for ISTEA in December 1991 was $155 billion for fiscal years 1992-1997.

The next major change came in federal funding when the Transportation Equity Act (TEA-21, Public Law 105-178) for the 21st Century was enacted in June 1998 at the end of ISTEA. TEA-21 provided a guaranteed level of $198 billion for transportation improvements for fiscal years 1998-2003. The theme of the funding was to rebuilding

American transportation system, increased safety, expanded opportunity for providing access to jobs, a cleaner environment for all and a balanced approach to highway programs. Table 19-1 shows TEA-21 funding allocation to Universities for transportation research.

Table 19-1 TEA-21 Funding for university research projects.

Year	1997*	1998	1999	2000	2001	2002	2003
	$ (M)	$ (M)	$ (M)	$ (M)	$ (M)	$ (M)	$ (M)
General Funds		6	1.2	1.2	1.2	1.2	1.2
Transit Account		0	4.8	4.8	4.8	4.8	4.8
Highway Account		25.65	25.65	25.65	25.65	25.65	25.65
Total	19.25	31.65	31.65	31.65	31.65	31.65	31.65

*ISTEA   Source: U.S. Transportation Department

Figure 19-2 shows the distribution of gas taxes from federal level to local levels. The distribution in general is based on formulas relating tot he local population.

State Funding

States have the responsibility of not only distributing federal funds, but also the funds collected by the state for transportation purposes. States get majority of their funding for transportation projects through retail sales of gasoline and diesel fuel, motor vehicle fuel licenses, motor vehicle weight fee, motor vehicle registration fees, driver license fee and miscellaneous taxes. These collected funds are transferred to Transportation Tax Fund (TTF) and State Transportation Fund (STF).

TTF funds are obtained from taxes on gasoline, diesel and vehicle license fees. Money from TTF funds is distributed into Motor Vehicle Fuel Account, Highway users Tax Account and Motor Vehicle License Fee Account. STF funds are obtained directly from the retail sales and user tax, motor vehicle registration and motor vehicle weight fees. Funds from STF are distributed into the Public Transportation Account, State Highway Account, Aeronautics Account, Bicycle Lane Account and Motor Vehicle Account.

To distribute these different funds, the states set up several programs such as State Transportation Improvement Program (STIP), Regional Improvement Program (handled by Counties), State Highway Operation and Protection Program and Congestion Management Program. In general there will be several programs for all transportation needs and areas.

Local agencies apply for project funding through STIP process. As part of this process even the state transportation department has to submit projects under STIP to

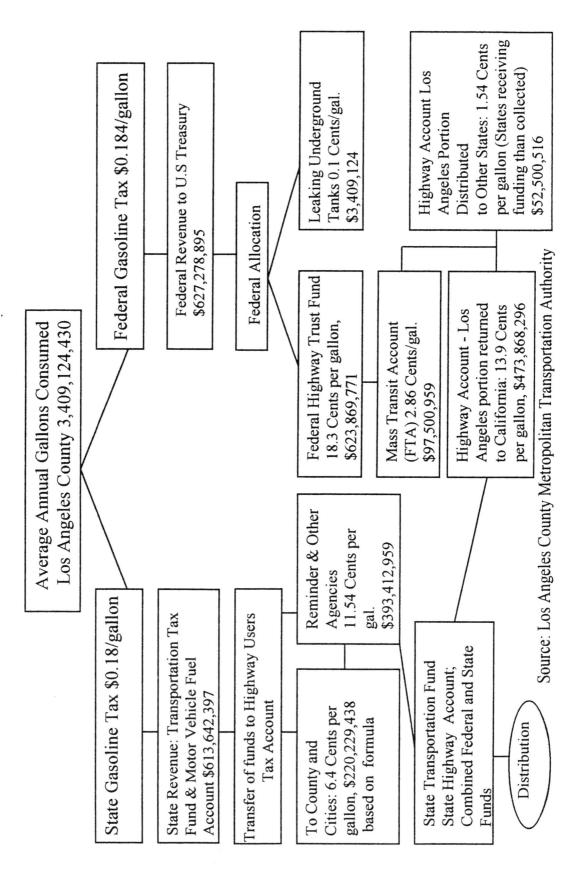

Figure 19-2 Sample federal and state gasoline tax distribution.

Average Annual Gallons Consumed Los Angeles County 3,409,124,430

Federal Gasoline Tax $0.184/gallon

Federal Revenue to U.S Treasury $627,278,895

Federal Allocation

Leaking Underground Tanks 0.1 Cents/gal. $3,409,124

Highway Account Los Angeles Portion Distributed to Other States: 1.54 Cents per gallon (States receiving funding than collected) $52,500,516

Federal Highway Trust Fund 18.3 Cents per gallon, $623,869,771

Mass Transit Account (FTA) 2.86 Cents/gal. $97,500,959

Highway Account - Los Angeles portion returned to California: 13.9 Cents per gallon, $473,868,296

State Gasoline Tax $0.18/gallon

State Revenue: Transportation Tax Fund & Motor Vehicle Fuel Account $613,642,397

Transfer of funds to Highway Users Tax Account

Reminder & Other Agencies 11.54 Cents per gal. $393,412,959

To County and Cities: 6.4 Cents per gallon, $220,229,438 based on formula

State Transportation Fund State Highway Account; Combined Federal and State Funds

Distribution

Source: Los Angeles County Metropolitan Transportation Authority

114

obtain funding for its projects. Once the state has determined under specific criteria and ranked the projects, it requests for federal funding. Similar to federal government, state government passes several legislation depending on local requirements to program and plan transportation projects. For example California has California Clean Air Act (1990), Transportation Development Act and California Environmental Equality Act (1970). The state government will also assist in developing Metropolitan Planning Organization and Metropolitan Transportation Authority based on size of the municipality. Funding is usually handled through strict guidelines that detail how the projects will be selected and funds distributed. To assist this process states prepare several types of manuals such as Project Development Procedure Manual and Local Program Guidelines. In providing the funding the states expect the various agencies to develop several planning documents. Most of the documents relate to long term planning for the region. The planning documents could relate to air quality, interregional travel, urban congestion relief and transportation development plan.

States also setup special funding such as Environmental Enhancement and Mitigation Demonstration (EEMM) fund within existing accounts. These funds will be used only for specific projects as defined by this account. For example EEMM funds are generally designated for landscaping and projects are ranked and selected for grant.

Local Funding

This section briefly describes how local funding is received and raised by counties, cities and special interest groups for transportation projects. The main source of funding for local agencies is federal and state funds. The funds available to local agencies are obtained making developing projects and applying for funds to federal, state and locally assigned metropolitan transportation agencies. Sometimes it becomes necessary to raise funds for local transportation needs. This is achieved in most counties by passing propositions for specific transportation improvement. For example, in Los Angeles County Proposition C, which imposes half-a-cent sales tax for all goods, sold within the county. The revenue is used to fund countywide transit-related improvements to freeways and state highways including countywide signal coordination and transportation system management projects. Other sources of local funds include private and public venture, advertising and leasing, benefit assessment districts, transit fare box revenue, developer impact fee and traffic violation fines. The special interest groups such as the Automobile Association, Transit Riders Union and Environmental Groups propose specific transportation related projects based on safety, mass transit and air pollution. They may obtain funding from federal, state and local agencies for their projects.

Transportation funding is a complex subject, very vast and difficult to compile. The different types of transportation projects and funding sources cause the complexity. A single large project could have several project phases with several funding sources. Also, it gets even more complicated when resources are pooled by different agencies for large projects. Metropolitan transportation agencies produce "funding sources matrix" which provide details about the source, description, eligible uses, polices and guidelines, annual spending level, project description and responsible agency.

In developing large projects it becomes necessary to pool the funding from many agencies, this sometimes brings institutional issues both technical and non-technical. Overcoming institutional issues sometimes is very difficult; it may involve back room meetings with all participants to agree on many issues. Funding for a project depends on policy issues, marketing, and stakeholder and equity issues. Also, it is required to consider interjurisdictional legal issues, private and public alliance, operation and maintenance issues and project performance criteria.

FIELD WORK AND DATA COLLECTION

In terms of fieldwork and data collection, tracking the type of project and the funding source in itself is a huge task. It is also important to know the limitations and jurisdictions within which to use the funds. The laws and regulations governing funding are complex and vary from project to project. Finally, the involvement of politics and politicians in allocating funds makes it even more challenging to keep track of funds.

FORMULATING CONCLUSIONS AND RECOMMENDATION

1. Determine how much your household spends on transportation needs on a monthly and yearly basis? List expenses and sources for automobile, transit and other transportation related (bikes, motor cycle and boats) uses. Discuss your findings.
2. Get your school budget and separate transportation related expenses, funding sources and limitations. The expenses to be listed are for parking, access roadways, lighting, security, maintenance facilities on campus, landscaping, insurance, operating budget for vehicles and personnel involved. Summarize your results in a table form.
3. From your research in question 2 are there any improvements that are necessary around your campus such as safety, lighting and parking needs. Determine the annual funds required for recommended improvements and sources for obtaining funding.
4. Request your local city or county or metropolitan transportation agency to provide you with a "funding source matrix." List projects that are receiving funding for projects in your area. If there is any transportation proposition (new or old) for raising funds through taxes for transportation projects passed by your state or county and obtain details (web page). Analyze the proposition and discuss whether it is needed (pros and cons)?

116

Lab 20

Intelligent Transportation System

STUDY OBJECTIVES

The main objective of this lab is to provide an introduction to Intelligent Transportation System (ITS) and to briefly explore the various areas of its application in transportation planning and traffic operations. Also, illustrate where the U.S. transportation system is heading in the 21st Century.

NEED FOR THE STUDY

As the demand for existing transportation network increases, the need to make the system efficient becomes a priority. In the U. S the number of vehicle miles traveled annually between 1985 and 1990 has increased nearly 60% (from 1.53 trillion miles to 2.42 trillion miles). The growth in public road system has increased during the same period by mere 1%. This has direct impact on congestion; the congestion in major metropolitan areas has increased from 7.3 million daily person-hours in 1982 to 14.2 million daily person-hours in 1993. The growth of suburban areas has been increasing and influencing the travel patterns. For example, the annual hour of delay and congestion cost per eligible driver in Los Angeles in 1996 is 44 and $1,205 respectively. Commuters are spending more and more time in their cars in trying to reach their destinations. Majority of the metropolitan areas including adjoining suburban areas are already built completely. That means there is very little or no space for new roads in most metropolitan areas. The overall demand has an impact on our environment leading to more pollution.

Under these pre-existing conditions, the question was how to make future transportation network efficient? One possible solution was to apply advanced technology such as computers, electronics, communications and management strategies to transportation and traffic engineering fields. Intelligent Transportation System (ITS) is the application of advanced technology in the fields of information processing and communications in the areas of traffic control, travel information, vehicle location and incident management. The potential benefit of using ITS is efficient use of existing and future transportation infrastructure. Also, improve safety, mobility, accessibility and productivity. It is expected that ITS will allow to operate and manage the transportation network as a single unit. This is achieved with the goal of developing inter-modal and multi-modal system across the country. One of the main objectives of the federal government is to reduce delay for all drivers by

117

fifteen minutes by the year 2010, that is savings of billions of hours annually leading to more productivity. Overall by improving the surface transportation there will be savings in energy and reduce pollution due to transportation.

GENERAL OVERVIEW

The potential for developing ITS technology and strategies was possible due to the passing of Intermodal Surface Transportation Efficiency Act (ISTEA) of 1991. The total budget for ISTEA in December 1991 was $155 billion for fiscal years' 1992-1997. A portion of this budget was used for initiating ITS. The next major investment from federal funding for ITS came when the Transportation Equity Act (TEA-21, for the 21st Century was enacted in June 1998 at the end of ISTEA. TEA-21 provided a guaranteed level of $198 billion for transportation improvements for fiscal years' 1998-2003. ITS act of 1998 provides funding for ITS infrastructure deployment and ITS Research and Development as shown in Table 20-1.

Table 20-1 TEA-21 funding for ITS projects.

Year	1998	1999	2000	2001	2002	2003	Total
	$(M)	$(M)	$(M)	$(M)	$(M)	$(M)	$(M)
Deployment	101	105	113	118	120	122	679
R & D	950	95	98.2	100	105	110	603
Total	196	200	211	218	225	232	1,282

Source. U.S. Transportation Department   M= Millions

Before ISTEA the transportation developmental plans and programs essentially focused on capital improvements with very little emphasis on maintenance and operation. The philosophy was that 'we could build out of congestion.' With the increasing demand and passing of ISTEA it became clear that old way of doing things was not sufficient to sustain existing transportation network. The new emphasis was improving operation, maintenance and increase cooperation between agencies. The new institutional cooperation was termed as multijurisdictional cooperation. This served as the basis for deploying ITS. This new cooperation allowed the to pool resources available to various agencies within a jurisdiction. Prior to ISTEA, ITS was used in a limited area such as for traffic signal coordination, automatic vehicle location (transit) and other areas. With ISTEA, ITS became part of transportation planning and investment decision making. Also, ISTEA emphasized that major metropolitan planning organizations and transportation agencies will be held accountable for implementing ITS and improving the environment.

ISTEA recognized the potential in implementing ITS and provided special research funds in order to "jump-start" the development of ITS technologies and strategies. This allowed the development of "National ITS Architecture." That is the development of a framework for implementing ITS across State, regional and local

118

jurisdictional boundaries and develop standards for ITS deployment across the country.

Some of the areas recognized for ITS applications in transportation planning and traffic operations fields are:

- **Highway and railroad safety study:** To improve warning systems, photo enforcement, traffic signal coordination and detect illegal vehicles crossing the rails.
- **Traffic Signal Control:** To improve signal system coordination, real-time data, emergency vehicles and transit vehicle priority during rush hours and development of Traffic Management Centers (TMC's). Use of advanced communication and information systems such as microwave, fiber optics, and video detection and spread spectrum radio.
- **Incident Management:** Development of systems on the roadway network for immediate response to incidents and thus reduces delay.
- **Travel Demand Management:** Includes development of travel information, ride-matching systems, automatic reservations, interactive media for kiosks, provide information on multi-modal travel and advanced systems that assist operators.
- **Electronic Toll Collection (ETC):** Technology that allows vehicles to go through without stopping at tolls. This includes closed circuit TV, video detection and two-way communications.
- **Emergency Response Management:** Similar to incident management to develop systems that assist automatically during emergency situations such as disasters and accidents. Also, develop systems to warn commuters about upcoming disaster or hazard.
- **Freeway and Arterial Management:** Efficient collections of traffic data, responds to needs on the freeways, improve variable sign messages and warn traffic on real-time basis.
- **Transit and Public Transportation Operation and Management:** Technologies that assist in efficient transit management. Provide two-way communications between vehicles, automated vehicle location technologies, improve performance of transit, systems that can assist in booking and provide assistance to travelers (kiosks). The goal is to improve operational performance of transit and make it attractive to non-users.
- **Commercial Vehicle Operation (CVO):** Development of automatic weighing machines, pre-clearance systems, electronic document clearance, two-way communication systems, linked computer systems to avoid delays at boarders and systems that assist in registration and credentials. Also, research is focused on fleet management of trucks, freight cars and rail cars for goods movement.
- **Advanced Vehicle Control and Safety System:** Development of Automated Highway System (AHS), collision avoidance system and travel map guidance system. Under AHS it is possible to platoon vehicles that are controlled electronically and thus increase lane capacity.

All the above areas use the transportation network over which ITS projects are overlaid. Therefore it is essential to evaluate ITS both qualitatively and quantitatively as to its contribution. The future of ITS is to include demonstration projects using ITS technologies in the Advanced Transportation Management System (ATMS) and Advanced Traveler Information System (ATIS). ATMS and ATIS are combination of above listed ITS projects. The application and deployment of ITS technologies are expected to fundamentally change surface transportation system. The most important element for ITS deployment is to achieve technical compatibility that is required between various systems being developed. What is termed as 'interoperability' between systems? For example the connection between field elements (Closed Circuit TV) and office elements (computers) should be possible when they are connected. The federal government is trying to ensure that field operational tests of ITS projects are compatible and reduces developmental costs when similar project is required to be deployed in another area. The National ITS Architecture is attempting to define key elements of ITS functions and data that must be exchanged between ITS subsystems for interoperability. ITS Research and Development portion are concentrating on Intelligent Vehicle Initiative, Metropolitan Travel Management, Rural ITS Services, Advanced, Advanced Public Transportation System and Commercial Vehicle Applications. Also, ITS research includes National Architecture development, ITS standards and developing technical assistance and training.

STUDY COMPONENTS

What's ITS? It involves the integration of electronics, communication, navigation, passenger information, and computer and control technologies into the transportation system. ITS is a new tool that is expected to increase mobility, increase accessibility, reduce energy use and provide cleaner environment. The introduction of ITS has changed the way traditional transportation planning being conducted by States and Metropolitan Transportation Agencies. It is important to study the changes brought by ISTEA to planning elements. The new planning process provides coordination, accountability, sharing of resources, multijurisdictional projects, joint decisions and management. Therefore there is a new breed of transportation planners trying to make ITS a success.

The next study component is to understand the impact of introducing advanced technology to transportation and traffic operations. Advanced technology such as the future traffic control systems, freeway management, incident management, electronic toll collection, emergency management and transit management. For example, the construction of a Transportation Management Center (TMC) can aid in monitoring and controlling traffic throughout the area plus provide assistance for transit monitoring, emergency management, and incident management. Figure 20-1 shows the various activities that could be conducted from a TMC.

The current focus of ITS is on standardizing and integrating systems that are deployed around the country at various ITS project locations. It is important to follow how ITS Architecture is developed for adaptation by manufacturers for achieving uniformity. In general it is necessary to understand technologies that make transportation more efficient encompass ITS. Some of the technology and methodology used may not be advanced or ITS oriented, but they incorporate ITS in one form or the other. For example, sharing of traffic volume and incident data between jurisdictions or allowing a truck to pass through after it has been registered in another jurisdiction. For ITS to be successful it requires sound management techniques and strategies. There are over 400 ITS projects underway or in process of being completed. It is important to link common transportation elements in a system and manage them as one unit. The same applies in financial management of projects, pool resources among jurisdictions for common benefit.

FIELD WORK AND DATA COLLECTION

ITS basically allows the gathering of information through advanced communication and delivers it to commuters and operators. In your region determine what projects are underway in ITS and group them under the six main categories listed below.

1. Travel and Traffic Management
2. Public Transportation Management
3. Electronic Payment
4. Commercial Vehicle Operation
5. Emergency Management
6. Advanced Vehicle Safety System

It will also be interesting to study the various advanced technology adopted for ITS. Currently, a broad range of diverse technologies such as information processing, communications, vehicle control, and electronics are being used. Combining these technologies is expected to provide products and service that makes transportation efficient. It is important to understand ITS standards and performance monitoring being developed. ITS standards are required to make the various systems interoperable. That is various systems can effectively work together and openly share information between jurisdictions. Also, interoperability and standards in manufacturing are expected to reduce cost and make system integration easy.

Make a visit to the nearest Transportation Management Center (TMC) in your area. List the various ITS project components used in the center for monitoring traffic.

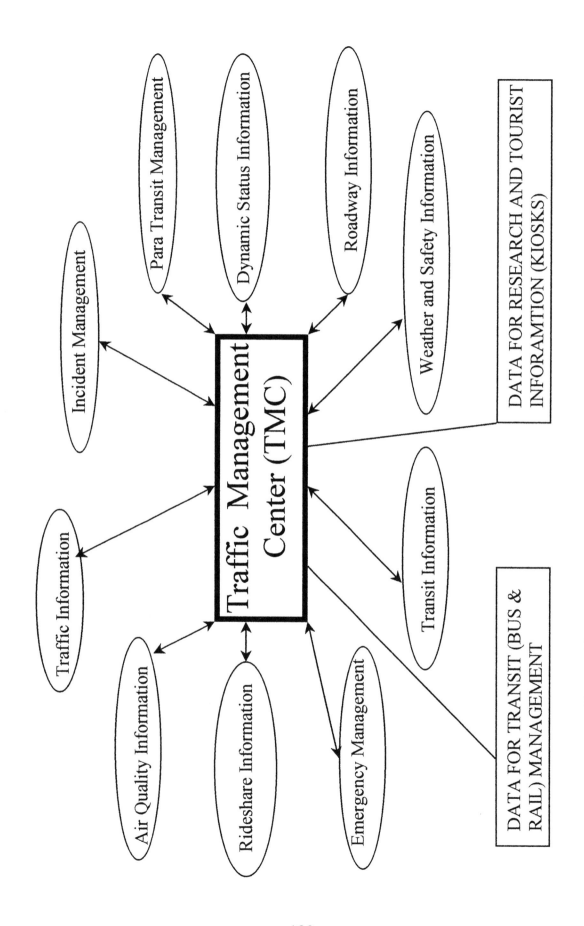

Figure 20-1. Basic functions of a traffic management center.

FORMULATING CONCLUSIONS AND RECOMMENDATIONS

1. Obtain from your state transportation agency or local metropolitan transportation agency a copy of ITS projects in your region? List the benefits from these projects.
2. What ITS applications have you observed on the roadways?
3. What technologies you think can be adopted to make transportation efficient and describe with an example?
4. Imagine a huge turnout of over 120,000 people for a sports event on a weekday during p.m. peak period. What ITS solutions would you recommend to alleviate traffic in and around the stadium? Describe.
5. What opportunities will ITS provided you in the job market? Discuss the areas and applications.

Lab 21

Transportation System Performance Monitoring

STUDY OBJECTIVES

The main objective of this lab is to provide students an understanding of what is Transportation System Performance Monitoring? How transportation projects can be evaluated? Discuss the importance and application of performance monitoring for various types of transportation planning and traffic operations projects.

NEED FOR THE STUDY

In everyday activity when you buy a product or get some kind of service you look for certain characteristics to evaluate what you are getting for your investment. Your evaluation could be either quantitative or qualitative or both. The same applies to your school education, after completing each course you get your transcript. A transcript is nothing but performance evaluation of your work in the classroom. Similarly transportation projects need to be monitored for their performance. Majority of transportation projects require large investments in terms of millions of dollars. Therefore performance-monitoring measures are like indicators of achievement. In transportation, performance measures are usually selected to measure levels of mobility, reliability and accessibility. Also, it includes customer satisfaction, cost-effectiveness, or progress towards stated goals and objectives of a project.

Why we need performance criteria? We need to measure the output of alternative projects, to rank projects for selection and combine projects to maximize output. The need arises because the public funds majority of transportation projects by paying tax for using the transportation system. When transportation projects are proposed and implemented, it is required to evaluate the benefits derived so the public will have faith in the system and its capabilities. The performance evaluations are written in technical memorandum that describes the benefits in simple and understandable language to the public. Performance criteria can be evaluated separately for short and long term phases of a transportation project.

Finally, the performance criteria of various projects in a region can be put together to see the regional benefits achieved due to initiating specific policies and projects. Also, performance criteria allow checking whether the initial goals and objectives of the project are met.

124

GENERAL OVERVIEW

In any system it is necessary to establish criteria against which the performance of a system can be measured. In case of a transportation system the criteria can be either quantitative or qualitative measures. Quantitative measures require specific type of data, for example measuring such items as changes in travel time, fuel consumption, vehicle occupancy, and The quantitative measures need large amount of data that was collected before the project and after the project for evaluation purposes.

It is not sufficient to gauge the complete success of a system. The total evaluation needs to include qualitative measures when it comes to transportation projects. A qualitative assessment like a survey about the safety of the system, comfort, attitude of passengers and overall system behavior provides a different view of the system. The qualitative measures provide the decision-maker, personnel involved, general public and management to make necessary changes to those elements that do not show up in quantitative analysis of the system. Qualitative measures provide information on service oriented measures and societal benefits. Another important element that needs to be considered is the external impact of transportation projects on the environment. These external impacts could be to measure noise impact, air pollution, wetland's impact, water quality, effect on animal migration and dislocation of business or households. In general good performance criteria will include both quantitative and qualitative measures as applicable to the project.

Transportation project such as building an arterial roadway or a bridge will have an output and an outcome. It is similar to quantitative and qualitative analysis. That is output of the system relates to the accomplishment of the goals of those building the roadway as the number of miles of new roadway added to the system. The outcome will try to measure the comfort, accessibility and mobility experienced by the users. Both output and outcome can be looked in terms of efficiency and effectiveness of the system respectively.

How should we use performance measures? In general it should be used only as a tool to gauge the efficiency and effectiveness of the project. The results obtained should be able to create information for future investments for decision-maker and not make it a political tool. The other major important thing to performance measures is to concentrate on the process of developing suitable criteria rather than on already perceived results due to the project. In most transportation projects the anticipated output and outcome may vary due to the dynamics of a place. Also, as most transportation projects are public funded it requires the involvement of public and all levels of government. The inclusion of public will allow understanding the perspective of the actual users through stakeholders meeting, public forums and general meetings across the jurisdiction.

The inclusion of performance-based evaluation for transportation projects allows the decision-maker to allocate project funding based on proven results. This allows greater accountability in investment of public transportation funds

STUDY COMPONENTS

There are various types of transportation projects that include both planning and operation elements. Some of the common projects are construction of freeways, high occupancy vehicle (HOV) lanes, arterial streets, signal synchronization (TSM), parking structures, ride share program (TDM), intelligent transportation system (ITS), bus and rail stations, landscaping and transit (rail and bus) improvement projects. The question is how do we measure the performance of each project individually as well as when projects are combined (Intermodal and Multi-modal) over a region.

For a complete evaluation of a project we need both quantitative and qualitative measures. We need to list the output as well as the outcome of transportation projects. Some have direct benefit to cost ratio, whereas some have either benefit or impact depending on the project type. For example when an arterial street is signal synchronized for about five miles, the benefit to the arterial can be calculated in terms of benefit to cost ratio. Whereas there is a negative impact in terms of additional delay to the side street users along the five mile section of the arterial. That is the side street users have a non-monetary (discomfort of waiting longer than before) and environmental (additional idling) and economical impact (excess gas usage). So, it becomes necessary to objectively consider all benefits and impacts and make trade-off as required between projects using proper value of judgment.

Table 21-1 shows the various performances measuring indexes that are used for a multimode system, Transit operating system (rail and bus) and Highway operations (freeway and arterials), Infrastructure Development and Transportation Demand Management projects. The performance measures shown in Table 21-1 relate to basic goals that are required for a transportation project. For example, mobility measure indicates the ability of people to go from place to place within reasonable amount of time. Accessibility relates to the information available to commuters using transit mode, particularly for those who are disabled. Air quality measure relates to the reduction of environmental impact due to the project, measured in terms of reduction of pollutants. Cost effectiveness measures the benefits in terms of financial performance of the transportation system. It could also be used to measure operating efficiency for transit mode. Quality and reliability measures indicate the true user benefits due to the project. A service provider has the obligation to provide clean and reliable service to the commuters.

Some of the data requirements for developing a performance evaluation procedure are first to determine existing transportation data. That is collect available data "before" the project is initiated. It is called as base line data. To conduct the performance evaluation, it is required to have same data sets that are collected "after"

126

the project is completed. This will allow measuring the performance of the project. The general types of data required in using Table 21-1 are:

1. Traffic Counts: Average Daily Traffic, Truck Volumes, HOV data, Lane capacity, Lane Volume, average vehicle occupancy (ridership), number of lanes and counts locations.
2. Congestion Data: Hours of delay, Travel time between origin and destination, speed, level of service, volume to capacity ratio, duration of congestion freeway and arterial, number of accidents, construction delays, emission reduction and fuel consumption.
3. Transit Data: Number of routes, frequency of buses/rail, number of buses, number of stops/stations, transit fare, boarding per day, miles of tracks, operating cost, number of hours of operation, maintenance cost, safety and security costs, vehicle miles of travel and average vehicle ridership by mode.

Using the above type of data it is possible to determine performance measures that are most understandable, meaningful and informative. Also, the data could be used to conduct sensitivity analysis. The sources of data could be state transportation offices, metropolitan agencies, county, city and private transportation consultants.

FIELD WORK AND DATA COLLECTION

Travel time-based measures are best suited for congestion measures. Travel related data could be collected to conduct simple congestion measures. Select a destination that is 10 miles away from your home. For example shopping mall or sports stadium or airport. As described in Lab 2 Travel Time and Delay Study, travel to your destination using auto by 1. Freeway 2. HOV 3. Arterial streets and 4. Combination.

Also travel to the same destination using local transit (bus and rail). During travel collect as much information as you can during the trip with respect to travel time, delay time, travel cost (approximation) and any other data you think helps you with performance evaluation. Conduct a quantitative and qualitative analysis of your trip using performance criteria listed in Table 21-1. Answer the questions under formulating conclusions and recommendations.

Table 21 Transportation projects performance measures

Type of Project	Mobility Measures	Accessibility Measures	Air Quality Measures	Benefit to Cost Ratio	Qualitative Measures	Reliability Measures
Multimodal Project	Measure of trip distance Average speed of modes Total commute time Average vehicle ridership	Number of transfers required Accessibility between modes	Air Quality Index (some measure of air quality improvement)	Cost effectiveness index. (benefit to cost ratio)	Consumer satisfaction Satisfaction (commuter survey) Safety and security	Schedule connections Response to customer needs
Transit System Operation	Daily Boarding Transit Mode Share Routing by corridor Frequency of trips	Employment accessibility (for persons without autos)	Percentage of fleet with alternative fuels.	Cost per passenger mile Cost per service hour Boarding per service hour Farebox recovery	Passenger load factor Average vehicle age distribution in fleet Consumer satisfaction	Passenger pass-ups On-time performance vehicles (fleet).
Highway Operations (freeway and arterials)	Freeway volume and speeds Truck volume on roadway Level of service Freeway and street delays Arterial street signal coordination data(TSM) HOV speeds and AVR	Auto ownership trends Population density at various locations Type of employment and business	Types of vehicles and characteristics of vehicles on roadway Types of trucks and percentage of alternate fuel capability (like CNG, Electric, etc.)	HOV utilization (a measure of construction cost of a HOV lane to the benefits due to travel time, savings, air quality etc.)	Measurement of pavement surface on roadway system. Percentage of roadway with deficiency	Response to freeway assistance Freeway incident duration Number of call-box use and call-boxes that are operational
Infrastructure Development (bridges/ramps)	People moving capacity before and after construction	Community served and environmental justification	Emission reduction potential	Benefit cost ratio Adherence ot budget Adherence ot schedule	Improvement of quality of life in the community. Any impacts to community	Safety measures required. Security measures required.
Transportation Demand Management Projects	Drive alone mode share Avg. Vehicle Occupancy Percentage travel by mode	Employment accessibility	Emission reduction potential	Cost/vehicle trip reduced Cost/vehicle miles travelled	Schedule availability and Location of stops	On-time performance Response to riders needs

Source: Los Angeles County Metropolitan Transportation Authority.   Note: Data needs to be collected for "before" and "after" conditions.

128

FORMULATING CONCLUSIONS AND RECOMMENDATIONS

1. What is the reduction in delay measured between freeway, HOV and arterial reaching the destination by auto? What performance measure can be used to distinguish the facilities used?
2. Compare the delay in reaching the destination using auto and transit? What is the estimated cost of travel (list all costs for both modes) considering reasonable cost of using auto and transit for your trip?
3. What are the qualitative performance criteria you could apply for your trip by auto and transit?
4. Do you think HOV is cost effective from your experience on your local freeways? Discuss.
5. What performance measure data can be collected for a signal coordination project in Lab 15?
6. Contact your local MPO and obtain a copy of performance measures used for various transportation projects? Discuss the methodology adopted.
7. What performance measures would you recommend for building a 10-mile bikeway in your community?
8. What performance measures would you recommend to evaluate Intelligent Transportation System used to provide signal coordination on an arterial street?

Lab 22

Transportation and Politics

STUDY OBJECTIVES

The main objective of this lab is to provide a general view of how politics plays a key role in selection, ranking and implementation of transportation projects in a region. To provide information on the organization structure of metropolitan planning organizations and other transportation related agencies (Federal, State, County and City) that have influence on shaping the transportation network and infrastructure in a region.

NEED FOR THE STUDY

Labs 1 through 21 provide the students with analytical methods and basic instructions on transportation planning and traffic engineering. There are very few details in these labs that tell the students how a transportation project actually comes to life and with whose blessing? It is interesting to know how, why and who affects a transportation project. This lab will provide a general idea as to the impact of politics on transportation planning process, that might not be available by just learning the analytical skills required.

Transportation today has been recognized as one of the most important element responsible for having good economical conditions in a region or country. Good transportation planning, policy and programs put forth by the decision-maker can develop a sustainable transportation network that is efficient. For example good transit systems can shape urban development and appropriate land uses. Another example is the development of suburbs around freeways. The need for mobility should be met by balanced land use and efficient transportation network. Also, the decisions made have an impact on quality of life within a region or country.

Many of the transportation projects cost millions of dollar. For example building a simple interchange for a freeway may cost between 5 to 10 million dollars. Majority of the large transportation projects are selected by politicians who are appointed as officials in transportation related committees. They may not have any direct experience in transportation. Therefore, understanding and affecting the political process is essential for developing a good regional transportation plan. Large transportation projects such as construction of bridges, roadways, railway lines, transit centers and freeways bring millions of dollars to a region and provide

130

employment to thousands of people. So, there is need for everyone in the community to understand the relation between transportation and politics.

GENERAL OVERVIEW

First it is important to understand what is planning? Planning is trying to mitigate anticipated problem in the future through some kind of guidance derived from previous experiences and current practices. Transportation planning in general involves the development of plans, policies and programs developed to improve the transportation in a region. A transportation planner or analyst may develop policies, project and programs to improve the transportation system in a region. The institutional and political environment binds transportation planners and analysts who work for governmental agencies. What about staff in a private consulting firm? They are not affected the same way as someone working for a governmental institution, but the institutional and environmental environment also influences them. In an institutional setting, the planner is challenged by constant ethical and political issues. This is mainly because of it involves public work and public funds. Any decisions made will have some kind of criticism, people who favor the decision and people who oppose the decision. Also, the question of social equity comes into picture with each policy and program developed. Did anybody benefit more in the community due to the policy and project? Is this project necessary? Is the data analyzed for this project valid? Does the project adhere to all legal requirements? Who was responsible for this project? All planning projects seem to face these questions.

Politics has a hand in everything we do on a daily basis. For example, specific laws are passed how your car should be manufactured; there are manuals that tell the traffic engineer what sizes the signs should be on a street; if you violate traffic there is a set fee; and where the next freeway segment should be built? Somewhere, a group of decision-maker is making decisions that govern our everyday life. This group of decision-maker could be elected appointed or hired personnel to do their jobs. We read news headlines such as the following:

> *"California Legislature Delays hearing on Controversial HOV Bill"*
> *"Federal Transit Authority Selects Projects for Federal Bus Rapid Transit Program."*
> *"Federal Highway Administration Issues Guidance on HOV Changes." and*

Governments are setup to provide protection to the public and provide essential services such as transportation, education and facilities and develop policies and programs as required. Majority of the decisions reached by government are based on technical, legal and political grounds. Technical aspects relate to data and necessary solution to the problem based on current practices for that particular problem or project. Legal aspects include the laws that are applicable to the project. The political aspect relates to how various groups of people in the community perceive the solution and how they will react to it. What does an official want? They would like to do

METROPOLITAN PLANNING AGENCY

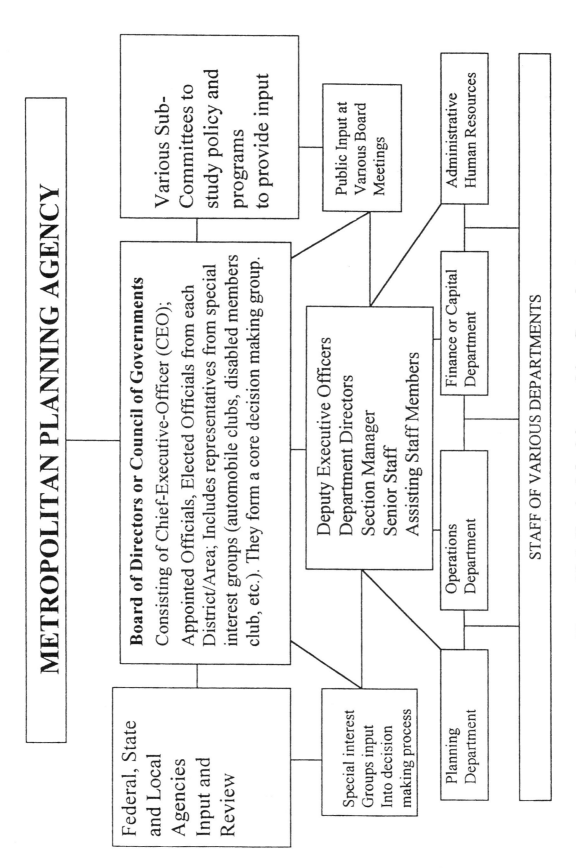

Board of Directors or Council of Governments

Consisting of Chief-Executive-Officer (CEO); Appointed Officials, Elected Officials from each District/Area; Includes representatives from special interest groups (automobile clubs, disabled members club, etc.). They form a core decision making group.

Various Sub-Committees to study policy and programs to provide input

Public Input at Various Board Meetings

Federal, State and Local Agencies Input and Review

Special interest Groups input Into decision making process

Deputy Executive Officers
Department Directors
Section Manager
Senior Staff
Assisting Staff Members

Administrative Human Resources

Finance or Capital Department

Operations Department

Planning Department

STAFF OF VARIOUS DEPARTMENTS

Figure 22-1. Typical decision making flow chart.

132

politically the right thing, so that they credit for their work. We know that some solutions however good will not be accepted by some sector of the population. Therefore, most major transportation projects have to be analytically sound, legally defensible and politically prudent. Figure 22-1 shows a typical organizational chart for a metropolitan planning organization. A policy, project or program has to grow from the staff level and move all the way to the Board of Directors for approval and implementation. Large projects take several months before being presented to the Board. The agency staff working below the Chief Executive Officer completes all the background work.

STUDY COMPONENTS

Who are these decision-makers? At governmental level decision-makers are elected officials, appointed officials and staff. In Figure 22-1, the elected officials of a region are usually appointed as Board of Directors for metropolitan planning organizations. Appointed officials serve at the request of the Chief Executive Officer, Mayor, Governor or the President. The appointed officials to the board include Board of Supervisors for the County, Executives from Mayors Office, Representatives from larger Cities within the metropolitan jurisdiction and representatives from Special Groups. Majority of these appointed officials will have some political goals to achieve such as becoming a Mayor, Governor, Senator, etc. They participate in making decisions that may bring them support in the form of votes from their local jurisdictions. That means, the elected or appointed official would support policy, project or program that they perceive as something that benefits them and benefits the public they serve.

How does the decision process work? A project will be approved only if consensus building has already taken place, the right political support is available and the economic climate is good. In general, the Board members with the help of staff work to develop projects that are analytically correct, legally binding and politically acceptable. Not only they initiate projects, they also assist in getting transportation funds from appropriate sources to their jurisdiction. If for some reason the appointed official has a legitimate reason not to support a project; the objections of the official will be heard and recorded for correction by the staff. Usually a Board member will listen to the public comments and to the person or agency they are representing. They make decisions based on analytical, legal and political impact of the project. Generally they support a project, policy or program that provides benefit to the public, demonstrates technical soundness and will not cause any loss politically to them or people they represent. For most meetings the public is invited, particularly to all board meetings. The public concerns are heard and documented by staff for action. The board members often debate public concerns and appropriate action is taken on each item and instructions given to the staff.

Who does all the real work? Public agency staff conducts the majority of the initial work. The staff is generally responsible to provide data, technical expertise, analyzes results, prepare memos, conduct meetings and interpret policies and guidelines. Majority of the time staff does initiate a project in a governmental organization and select suitable consultant to finish and implement the project. For most projects the staff determines the solution that is appropriate and supports the organization position. Chief Executive officer and appointed officials are responsible for setting broad policies that govern the project direction and implementation. It is required by the staff to interpret the policies and determine appropriate solution.

Will your solution be accepted? It all depends on how much groundwork was done before it was taken to the Board for decision. It depends on several factors including the decision maker, lobbyists and current political support (public) for the project. However technically sound your project may be, sometimes the project will not be approved during a board meeting due to politics. As a staff member you should be ready to digest such reversals.

What happens in the County and City levels? At County and City offices, the same methods are adopted in developing a project, but will be more internal. Each County and City depending on the size has separate transportation departments. Elected Board of Supervisors mostly controls the County affairs. At a City level it is the Mayor, City Manager and City Council Members. The County, City, State Transportation Department and Special Groups attend meetings conducted by the Metropolitan Planning organizations and provide input in developing projects.

Once projects are listed a Regional Transportation Improvement Plan (RTIP) is developed with the approval of all agencies and sent to the state offices for funding. The state in turn will collect projects from various metropolitan planning organizations and prepare a State Transportation Improvement Plan (STIP). At state level the same procedure as in a metropolitan planning organization is carried out in a different manner. At state political influence plays a role in ranking and allocating funds for projects. The STIP is sent to federal offices (FHWA and FTA of U. S. Department of Transportation) for evaluation and funding. At federal level the political influence is much higher, the merits of the project and the amount of money to be allocated will be debated. Overall politics does play a role in transportation planning, but the amount or level of play depends on the size of the project, the place where the project originated and the political clout the jurisdiction has on rest of the decision-makers.

FIELD WORK AND DATA COLLECTION

Obtain organization charts for your Metropolitan Planning Organization, County Public Works Office and City Transportation Department. Also, obtain a list of committees, sub-committees and names and titles of decision makers in each organization.

FORMULATING CONCLUSIONS AND RECOMMEDNATIONS

1. List your County Board of Supervisors, City Council Members and State Senators? Determine is any of them are serving on any of the transportation related agency or committees?
2. Determine a freeway project (or any large transportation project) in your area and find out which agency was responsible in initiating the project; which politician was responsible for obtaining the funds and how much funding was obtained?
3. Pick current newspaper and determine any transportation related policy or project that is being debated in your State, County or City? Discuss the players (jurisdiction, benefits, impacts, location of project, politician, party affiliation etc.) in it?

Appendix A

Transportation Journals and Library
and
WWW Sites and Job Sources

Institute of Transportation Engineers Journal; Published by Institute of Transportation Engineers, Washington D.C. www.ite.org. Telephone 202-554-8050.
Subject Area Covered: Transportation planning, traffic engineering and transportation and traffic safety in regard to the movement of people and goods on the surface transportation system.

Intelligent Transportation Systems (ITS) Journal: Published by Gordon and Breach Science Publishers, Amsterdam, Netherlands. www.gbhap.com, Telephone 1-800-545-8398 (U.S.).
Subject Area Covered: Research, development, operational tests and deployment of ITS concepts and technologies; rigorous scholarly discourse on significant ITS issues.

ITS Quarterly: Published by ITS America, Washington D.C. www.itsa.org, Telephone 202-484-4847.
Subject Area Covered: Technical, Social and legal issues presented by ITS including new institutional arrangements between government and the private sector, protection of privacy, liability implications and effects on environment and energy.

Journal of Advanced Transportation: Published by Institute for Transportation Inc., Calgary, Canada. Telephone 403-286-9429.
Subject Area Covered: Oriented to urban mass transit, emphasizing advanced transportation concepts, ITS issues, planning and design issues and new technologies.

Journal of Transportation Engineering: Published by the American Society of Civil Engineers, Virginia. www.pubs.asce.org, Telephone 1-800-548-2723.
Subject Area Covered: Planning, design, construction, maintenance and operation of air, highway and urban transportation, as well as the pipeline facilities for water, oil, and gas. Specific topics include airport and highway pavement maintenance and performance, management of roads and bridges, traffic management technology, construction and operation of pipelines, and economics and environmental aspects of urban transportation systems.

Mass Transit: Published by Cygnus Publishing, Fort Atkinson, Wisconsin. www.masstransitmag.com, Telephone 920-563-6388.
Subject Area Covered: National and International issues in regard to bus, rail, and Intermodal transportation in the public and private transit industry.

Traffic Engineering and Control: Published by Printerhall Limited, London, England. Telephone 44-171-636-3956.
Subject Area Covered: All aspects related to traffic engineering and road safety.

Transportation Quarterly: Published by ENO Transportation Foundation Inc. Washington D.C. www.enotrans.com, Telephone 202-879-4700.
Subject Area Covered: All modes with focus on policy and planning.

Transportation Research Record: Published by Transportation Research Board (TRB), National Research Council, Washington D.C. www.nas.com, Telephone 202-334-3214.
Subject Area Covered: Collections of papers on specific transportation modes and subject areas. The series primarily contains the more than 600 papers prepared for presentation at the TRB annual meetings.

Transportation Science Journal: Published by Transportation Science Section, Operations Research Society of America, Baltimore, Maryland. www.informs.org, Telephone 410-850-0300.
Subject Area Covered: Topics include traffic engineering, science, vehicle routing, network equilibrium and location modeling.

Transportation Planning and Technology: Published by Gordon and Breach Science Publishers, Amsterdam, Netherlands. www.gbhap.com/Transportation_Planning_Technology, Telephone 1-800-545-8398 (U.S.).
Subject Area Covered: Academic/scholarly production.

Journal of Transportation and Statistics (JTS) : Published by U.S. DOT, Washington D.C. www.bts.gov/programs/jts, Telephone 202-366-DATA.
Subject Area Covered: The JTS provides a forum for the latest developments in transportation information and data, theory, concepts and methods of analysis relevant to tall aspects of the transportation system.

Public Transport International: Published by International Union of Public Transport, Brussels, Belgium. Www.uitp.com, Telephone 32-2-6736-100.
Subject Area Covered: All issues concerning public transport, modes of transportation, urban planning, traffic policies, data processing, human resources etc.

Related Journals: Journal of Transport Economics and Policy
World Transport Policy and Practice
Transportation Research Part A - Policy and Practices
Transportation Research Part B –Methodological
Transportation Research Part C – Emerging Technology
Transportation
Transportation Law Journal
Public Roads
Traffic Technology
ITS World
Highway and Transportation

Source: "Transportation Journals." The Urban Transportation Monitor, December 4, 1998.

WWW Sites and Job Sources

Name of Site	URL	Reasons for being useful
Institute of Transportation Engineers	www.ite.org	Articles, publications etc.; practice-oriented articles; up-to-date technical inf.; nationwide participation; good discussion forum; job advertising.
Bureau of Transportation Statistics	www.bts.gov	Reports and statistics; urban statistics; up-to-date technical information; data; comprehensive statistics; breadth of information; online library.
US Department of Transportation	www.dot.gov	Search engine; publications; regulations and guidelines; federal perspective; general information; current information; access to other federal sites.
Federal Highway Administration	www.fhwa.dot.gov	Policy and planning issues; most information; databases; guidelines; regulations; reports; what's new; MUTCD Editions.
Federal Transit Authority	www.fta.dot.gov	Good information on projects; budgeting and funding information; news and documents.
American Public Transit Association	www.apta.com	National lobbying; peer network; technical support; news; documents.
Transportation Research Board	www.nas.edu/ trb	Publication index/search function; recent reports; up-to-date technical information; ITS related information.
California Department of Transportation	www.dot.ca.gov	California Transportation inf.; contacts.

WWW Sites and Job Sources

Name of Site	URL	Reasons for being useful
McTrans	http://mctrans.ce.ufl.edu	Software news
American Association of State Highway and Transportation Officials	www.aashto.org	Current and relevant inf.
Urban Analysis Group	www.uag.org	VIPERT and trans. inf.
Environmental Protection Agency	www.epa.gov	Air Quality Documents
American Planning Association	www.planning.org	Links to MPO's, APA.
Florida Department of Transportation	www.dot.state.fl.us/planning/	Florida Department of Transportation Information
Texas Transportation Institute	http:/tti/tamu.edu	Mobility studies; well-organized site.
Washington State Department of Transportation	www.wsdot.wa.gov	Current information and contracts download software/manuals.
ISTEA	www.istea.org	Update on Federal Trans. Issues.

Source: The Urban Transportation Monitor, November 20, 1998
Note: All web sites last checked in February 2001.

Appendix B

Computer Software Applications
in
Transportation and Traffic Engineering

INTRODUCTION

Use of computers is assisting transportation professionals to be efficient and complete projects on time. There are many mathematical formulas and calculations that need to be done and some needs to be repeated, the role of microcomputers becomes very apparent in those situations. Now, with computers available to most professionals and with the advent of Internet information regarding transportation is available any time. The lap top computers and hand held mini computers have assisted in developing software that helps in data gathering and transferring as requited. As advances in microcomputers are growing, existing applications are being expanded. Also, interfacing different systems and software has given increased capability. For example both PC and UNIX are combined conduct to run complex travel forecasting models. Software from both systems is interfaced to obtain required output.

Presently there are over 300 transportation planning and traffic operational software applications available on the market. A majority of these software products can be obtained either from McTrans, Center for Microcomputer in Transportation, Transportation Research Center, University of Florida or PC Trans(see Appendix B). The programs are available through McTrans are programs created by individuals and agencies, and are useful in conducting planning, operations and management of transportation related projects.

The software applications available through McTrans are classified broadly as follows:

- **Highway Engineering:** Construction Management, Highway Design, Hydraulics, Pavements and Surveying.
- **Traffic Engineering:** Capacity Analysis, Data Processing, Demand Modeling, Network Assignment, Project Management and Site Impact Analysis.
- **Transportation Planning:** Data Processing, Transportation Demand Modeling, Network Assignment, Project Management and Site Analysis.
- **Transit:** Operations, Planning, Scheduling, Analysis and Publications.
- **General Interest:** Administration and Miscellaneous.

141

Mainframe version of several highway and traffic engineering programs also are available from McTrans. The prices of software and license fees vary depending on the software. There is a 50% discount available for universities.

SELECTION AND APPLICATION OF SOFTWARE

Choosing the right software depends on the specific problem at hand and required output. A preliminary check on the input data required for the software should be explored before buying the software. Also, it is necessary to collect all the required field data before using the software. Data requirements may vary depending on the software application. For example, the level of service (LOS) in Lab 8 can be done using traffic engineering software such as CAPSSI (Comprehensive Analysis Program for Singly Signalized Intersection) and HCS software based on Highway Capacity Manual. The filed data collected in Lab 8 can be used to determine the LOS. Using CAPSSI will allow you to do analyze one intersection at a time, whereas HCS software will allow the analysis of several intersections at one time. So, the software you need will depend on the job at hand and the type of output required. In the above example, the field data required are the turning movement counts, saturation flow rate, number of lanes (geometric) and information on signal phasing at the intersection. Both CAPSSI and HCS determine the LOS using above field data. For quick review of LOS CAPPSI preferable and for complete analysis of several intersections HCS software is essential. Both could be used to verify the results and variations if any.

Other software that is useful in conducting lab work in this book is, TCDI (Traffic Control Devices Inventory), SIS (Sign Inventory System), SIDRA (level of service at intersections) and TDPI (Traffic Data Input Program). For conducting signal warrants and signal timing analysis software such as PASSERIV and TRANSYT 7F can be used. A catalog of software and McTrans publications can be requested from McTrans (http://mctrans.ce.ufl.edu/). Visit various web sites shown in appendices and obtain free software from FHWA and FTA sites.

REFRENCES
McTrans. Center for Microcomputers in Transportation. Transportation Research Center, University of Florida, 512 Weil Hall, P.O Box 116585, Gainesville FL 32611-6585.
Microcomputer Applications in Transportation II. 1987. Proceedings of the North American Conference on Microcomputers in Transportation. American Society of Civil Engineers, New York.
Microcomputers Applications in Transportation III: Volume I & II. 1989. Proceedings of the International Conference on Microcomputers in Transportation. American Society of Civil Engineers, New York.
Microcomputers Applications in Transportation. 1992. Proceedings of the 4th International Conference on Microcomputers in Transportation. American Society of Civil Engineers, New York.

Bibliography

Lab 1

Manual on Uniform Traffic Control Devices for Streets and Highways. 1988. U.S. Department of Transportation. Federal Highway Administration, Washington, D.C.

Transportation and Traffic Engineering Handbook. 1982. Institute of Transportation Engineers, Washington, D.C.

Manual on Traffic Engineering Studies. 1976. Institute of Transportation Engineers, Washington, D.C.

Lab 2

Transportation and Traffic Engineering Handbook. 1982. 2nd Edition. Institute of Transportation Engineers, Washington, D.C.

Manual of Traffic Engineering Studies. 1976. Institute of Transportation Engineers, Washington, D.C.

Adolf. D. May. 1990. Traffic Flow Fundamentals. Prentice-Hall, Inc., New Jersey.

Lab 3

Paul H. Wright, and Randor J. Paquette. 1987. "Highway Engineering," 5th Edition. John Wiley and Sons, Inc.

Homburger, Wolfgang S. and James H. Kell. 1988. "Fundamentals of Traffic Engineering," 12th Edition. Institute of Transportation Studies. University of California, Berkeley.

Clarkson H. Oglesby. 1985. "Highway Engineering" , 3rd Edition. John Wiley and Sons, Inc.

Lab 4

Highway Capacity Manual. 1985. Special Report 209. Transportation Research Board. Washington, D.C.

Paul H. Wright, and Randor J. Paquette. 1987. "Highway Engineering," 5th Edition. John Wiley and Sons, Inc.

Bibliography

Manual of Traffic Engineering Studies. 19767. Institute of Transportation Engineers, Washington, D.C.

Lab 5

Paul H. Wright, and Norman J. Ashford. 1989. Transportation Engineering, Planning and Design, 3rd Edition. John Wiley and Sons, Inc. New York.

Manual of Traffic Engineering Studies. 1976. Institute of Transportation Engineers, Washington, D.C.

Transportation and Traffic Engineering Handbook. 1982. 2nd Edition. Institute of Transportation Engineers, Washington, D.C.

Lab 6

American Association of State Highway and Transportation Officials. 1990. "A Policy on Geometric Design of Highways and Streets." Washington, D.C.

Highway Capacity Manual. 1985. Special Report 209. Transportation Research Board. National Research Council, Washington, D.C.

Lab 7

Transportation Research Board, Highway Capacity Manual, National Research Council, 4th Edition, 1985.

Paul H. Wright, and Randor J. Paquette, Highway Engineering, John Wiley & Sons, 5th Edition, 1987.

Adolf D. May, Traffic Flow Fundamentals, Prentice Hall Inc., 1990.

Lab 8

Highway Capacity Manual. Special Report 209. Transportation Research Board, Washington, D.C., 1985.

Paul H. Wright and Randor J. Paquette, Highway Engineering. 5th Edition. John Wiley and Sons, 1987.

Bibliography

Lab 9

Donald Gross and Carl M. Harris, Fundamentals of Queuing Theory, 2nd Edition, John Wiley and Sons, 1985.

Daniel L. Gerlough and Matthew J. Huber, Statistics with Applications to Highway Traffic Analyses. ENO Foundation for Transportation, Inc. Westport, Connecticut, 1978.

Paul H. Wright and Randor J. Paquette. Highway Engineering, 5th Edition, John Wiley and Sons, New York, 1979.

Lab 10

Vergil A. Stover and Frank J. Koepke, Transportation and Land Development, Institute of Transportation Engineers, 2nd Edition, 1988.

Donald Gross and Carl M. Harris, Fundamentals of Queuing Theory, 2nd Edition, John Wiley and Sons, 1985.

Lab 11

Paul C. Box, and Joseph C. Oppenlander, Manual of Traffic Engineering Studies, Institute of Transportation Engineers, 4th Edition, 1976.

Vergil A. Stover, and Frank J. Koepke, Transportation and Land Development, Institute of Transportation Engineers, 2nd Edition, 1988.

Institute of Transportation Engineers (ITE), Transportation and Traffic Engineering Handbook, 2nd Edition, 1982.

Robert A. Weant, and Herbert S. Levinson, Parking, ENO Foundation, 1990.

Lab 12

Homburger ,Wolfgang S. and James H. Kell. Fundamentals of Traffic Engineering. 12th Edition. Institute of Transportation Studies. University of California, Berkeley, 1988.

C. John Khisty. Transportation Engineering on Introduction. Prentice Hall, Englewood Cliffs, New Jersey, 1990.

Bibliography

Institute of Transportation Engineers. Transportation and Traffic Engineering Handbook. 2nd Edition. Prentice Hall, Inc., Englewood Cliffs, New Jersey, 1982.

Lab 13

Barton-Aschman Associates and Cambridge Systematic, Inc.1997. "Model Validation and Reasonableness Checking Manual." U.S. Department of Transportation, Federal Highway Administration.

IBI Group. 1979. "Canadian Urban Modeling: A Review of Innovative Urban Modeling Techniques." Transport Canada, Surface.

U.S. Department of Transportation, Federal Highway Administration. 1973. "Calibrating & Testing a Gravity Model for Any Size Urban Area."

U.S. Department of Transportation, Federal Highway Administration. 1998. "Introduction to Urban Travel Demand Forecasting."

The Urban Analysis Group, Inc.1998. "User Manuals for Tranplan and NIS." Version 9.0

Transport Canada. 1979. "Canadian Urban Modeling: A Review of Innovative Urban Modeling Techniques." Prepared by IBI Group for the Urban Transportation Research Brach of Canadian Surface Transportation Administration.

Lab 14

"Ridership Still Growing."1999. Passenger Report, Volume 57, Number 30.
U.S. Department of Transportation, Federal Highway Administration. 1998. "Introduction to Urban Travel Demand Forecasting."

The Urban Analysis Group, Inc.1998. "User Manuals for Tranplan and NIS." Version 9.0.

Transport Canada. 1979. "Canadian Urban Modeling: A Review of Innovative Urban Modeling Techniques." Prepared by IBI Group for the Urban Transportation Research Brach of Canadian Surface Transportation Administration.

Lab 15

Evaluation of Travel Demand Management Work Program Report. 1993. Federal Highway Administration.

Bibliography

U.S. Department of Transportation. 1993. "Implementing Effective Travel Demand Management Measures: Inventory measures and Synthesis of Experience."

U.S. Department of Transportation.1993. "An Assessment of Travel Demand Management Approaches at Suburban Activity Centers."

Lab 16

Before and After Study Guidelines for Signal Synchronization Projects. Los Angeles County Metropolitan Transportation Authority. November 1997. Los Angeles. California.

Homburger, S. Wolfgang, et al., 1996. Fundamentals of Traffic Engineering. 14th Edition Institute of Transportation Studies, Berkeley, U.S.A.

Institute of Transportation Engineers (ITE). 1982. Transportation and Traffic Engineering Handbook. 2nd Edition. Washington, D.C.

Murthy, Narasimha A.S and Cude, Kenneth. 1998. "Quantifying Countywide Benefits and Impacts of TSM Projects." ASCE Conference on Transportation Land Use and Air Quality, Portland, Oregon.

Murthy, Narasimha A.S and Liu, Peter.1997. "Regional Traffic Forums: A Successful Tool for Implementation of TSM in Multijurisdictional Projects." The XIII International Road Federation World Meeting, Toronto, Canada.

Signal Synchronization and Bus Speed Improvements Program Guidelines. Los Angeles County Metropolitan Transportation Authority. February 1995. Los Angeles. California.

Lab 17

U.S. Department of Transportation, Federal Highway Adminstration.1985. "Site Impact Traffic Evaluation (S.I.T.E) Handbook."

Institute of Transportation Engieers.1991. "Traffic Access and Impact Studies for Site Development: A Recommended Practice."

Institute of Transportation Engieers.1995. "Trip Generation Handbook."

Institute of Transportation Engieers.1995. "Transportation and Traffic Engineering Handbook."

Bibliography

Murthy, Narasimha A.S. 1992. "Traffic Data Collection: What Really Needs to be Done?" ASCE Site Impact Traffic Assessment Conference, pp. 1 - 5, Chicago, Illinois.

Lab 18

"Before and After Study Guidelines for Signal Synchronization Projects." 1997. Los Angeles County Metropolitan Transportation Authority. Los Angeles. California.

"Final 1994 Air Quality Management Plan: Meeting the Air Quality Challenges." 1994. Southern California Air Quality Management District. Diamond Bar, California.

Griffin, Roger D. 1994. "Principles of Air Quality Management." Lewis Publishers, CRC Press Inc.

Murthy, Narasimha A.S and Cude, Kenneth. 1998. "Quantifying Countywide Benefits and Impacts of TSM Projects." ASCE Conference on Transportation Land Use and Air Quality, Portland, Oregon.

Murthy, Narasimha A.S. 1995. "Transportation Control Measures to Reduce Air Pollution." Transportation Conference, The Nevada Chapter of ASCE & ITE, Las Vegas, Nevada. "Transportation Planning and Air Quality." 1992. Proceedings of the National Conference sponsored by the American Society of Civil Engineers.

Lab 19

California State Board of Equalization.1998. Taxable Sales In California (Sales & Use Tax): Third Quarter.

Institute of Transportation Engineers (ITE). 1982. Transportation and Traffic Engineering Handbook. 2nd Edition. Washington, D.C.

U. S. Department of Transportation. 1999. Financing Federal-Aid Highways. Publication No. FHWA-PL-99-015. Office of Legislation and Strategic Planning, Washington D.C.

A Summary Transportation Equity Act for the 21st Century: Moving Americans into the 21St Century. 1998. U. S. Department of Transportation.

Bibliography

How To Keep America Moving ISTEA: Transportation for the 21st Century. 1997. U. S. Department of Transportation.

U.S. Department of Transportation. 1992. A Guide to Federal-Aid Programs, Projects, and Other Uses of Highway Funds. Publication No. FHWA-PD-92-018. Federal Aid Program Branch, Washington D.C.

Signal Synchronization and Bus Speed Improvements Program Guidelines. Los Angeles County Metropolitan Transportation Authority. February 1995. Los Angeles. California.

Tadi, Ramakrishna and Murthy, Narasimha. 1997. "An Innovative Approach to Transportation Infrastructure Financing." The XIII the International Road Federation World Meeting, Canada.

"Slants & Trends," Urban Transport News, Vol.27 No.7, Page 49, April 7, 1999.

"Tea-21 Provisions Pose Problems for Appropriators, Grantees." Washington Letter on Transportation, Vol. 17, No. 34, page 1, April 24, 1998.

Lab 20

A Summary Transportation Equity Act for the 21st Century: Moving Americans into the 21st Century. 1998. U. S. Department of Transportation.

How To Keep America Moving ISTEA: Transportation for the 21st Century. 1997. U. S. Department of Transportation.

U.S. Department of Transportation. Transportation Planning and ITS: Putting the Pieces Together. Federal Highway Administration. Washington D.C.

Lab 21

De Coral - Souza, Patrick, Everett, Jerry, Gardner, Brian et al. 1997. Total Cost Analysis: An Alternative to Benefit-to-Cost analysis in Evaluating Transportation Alternatives. Transportation Journal, Vol. 24, pp. 107-123. Netherlands.

Higgins, J. Thomas. Fall 1995. Congestion Management Systems: Evaluation Issues and Methods. Transportation Quarterly, Vol. 49, No. 4, pp. 25-42.

Los Angeles County and Ventura Counties ITS Strategic Deployment Plan. September 1997.

Bibliography

Stuart, G. Darwin. Winter 1997.Goal Setting and Performance Measurement in Transportation Planning and Programming. Journal of Public Transportation, pp. 49-71.

Technical Memorandum Performance Criteria. National Engineering Technology Corporation.

The Urban Transportation Monitor. August 1997. Significant New Study Evaluates Congestion Measures.

Lab 22

"California Legislature Delays hearing on Controversial HOV Bill." The Urban Transportation Monitor, April 30,1999.

"FTA Selects Projects for Federal Bus Rapid Transit Program." "California Legislature Delays hearing on Controversial HOV Bill." The Urban Transportation Monitor, June 25,1999.

"FHWA Issues Guidance On HOV Changes." The Urban Transportation Monitor, August 20,1999.

"San Diego MTDB Adopts Transit/Land Use Planning Coordination Policy." The Urban Transportation Monitor, August 20,1999.

Forester, John.1989. Planning in the Face of Power. University of California Press, Berkeley.

Klein, D. Richard. 1990. Everyone Wins: A Citizen's Guide to Development. Planner Press, American Planning Association, Washington D.C.

Lewis, C. Nigel. 1993. Road Pricing. Thomas Telford Services Ltd., Telford House, London.

U.S. Department of Transportation. 1995. Corridor Preservation: Study of Legal and Institutional Barriers, Federal Highway Administration, Washington D.C.

Glossary of Terms

ALL –OR –NOTHING ASSIGNMENT – The most basic form of traffic assignment which loads all trips, for each zone pair to one shortest path.

APPROACH – A set of lanes accommodating all left-turn, through and right turn movements arriving at an intersection.

CAPACITY – Capacity of a facility is defined as the maximum hourly rate at which persons or vehicles can reasonably be expected to traverse a point or uniform section of a lane or roadway during a given period under prevailing roadway, and traffic control conditions.

CENTRAL BUSINESS DISTRICT (CBD) – Usually the busiest downtown retail area of a City. High land values and large volumes of traffic are typical in a CBD.

CENSUS TRACT – A small statistical subdivision of a country. Generally tracts have stable boundaries of about 2500 to 8000 residents.

CENTROIDS – Centroids identify the center of activity within a zone and connect that zone to the transportation network.

CENTROID CONNECTOR - Centroid connectors represents the path from a centroid to the highway or transit network. They usually represent local streets that carry traffic into the TAZ.

CENTROID CONNECTOR – The links that physically connect the centroids to the network. Most of the time they are thought of as local streets.

CHANGE INTERVAL – The "yellow" plus "all red" intervals that occur between phases to provide for clearance of the intersection before vehicles from conflicting directions are released; stated in seconds, and given the symbol Y.

CONTROL CONDITIONS – Refer to the types and specific design of control devices and traffic regulation present on a given facility. The locations, type and timing of traffic signals are critical control conditions affecting capacity. Other important control devices include STOP and YIELD signs, lane use restrictions, turn restrictions, and similar measures.

CORDON LINE – The boundary for the study area.

Glossary of Terms

CYCLE – Any complete sequence of signal indications.

CYCLE LENGTH – The total time for the signal to serve all phases, stated in seconds, and given the symbol C.

DELAY – Additional travel time experienced by a driver, passenger, or pedestrian beyond what would reasonably be required for given trip.

DENSITY – The number of vehicles occupying a given length of a lane or roadway, averaged over time, usually expressed as vehicles per mile (vpm).

ELASTICITIES – Transportation elasticities are the ratio of change in travel demand in response to changes in transportation system.

FRICTION FACTOR – Friction factor represents the impedance of travel between an origin and destination. The measures could be travel time, cost or distance.

FREEWAY – A facility with full control access, giving preference only to through traffic.

EXTERNAL STATION – These are located on the major transportation routes into/out of the study area.

FORECAST YEAR – In modeling it refers to the future year for which forecast is done to determine the number of highway and transit users.

FLOW RATIO – For a given approach or lane group; the ratio of the actual flow rate for approach or lane group, v, to the saturation flow rate. The flow rate is given the symbol, (v/s), for approach or lane group i.

FULLY ACTUATED OPERATION – In a fully actuated operation, all signal phases are controlled by detectors.

GEOGRPHIC INFORMATION SYSTEM (GIS) – An integrated data storage and output program, comprised of a pure database linked to a graphic display unit. It is designed to capture, store, retrieve, analyze, and display data files.

GRAVITY MODEL – Adapted from Newton's Law of Gravitational Force between Masses. In using this model for transportation forecasting it is required to have relative attractiveness of a zone as a measure of impedance of a zone.

GREEN RATIO – The ratio of effective green time to the cycle length; given the symbol g/C (for phase I)

Glossary of Terms

GREEN TIME – The time within a given phase during which the "green" indication is shown, stated in seconds and given the symbol G, (for phase I).

HEADWAY – The time between two successive vehicles in a traffic lane as they pass a point on the roadway, measured from front bumper to front bumper, in seconds.

HIGHWAY PERFORMANCE MONITORING SYSTEM (HPMS) – A HPMS system helps to track information of a large area such as a State's traffic volume, vehicle classification and other travel and transportation related information.

HIGH OCCUPANCY VEHICLE (HOV) LANES – Lanes permitting only more than one driver to be in the vehicle. Some HOV lanes require more than two people to be in the vehicle for use.

HOME-BASED WORK – a trip whose origin is at home and ends at home after completing the work.

IMPEDANCE – A measure of time and/or cost expressed in terms of the generalized cost of traversing a particular link.

INTELLIGENT TRANSPORTATION SYSTEM (ITS) – A collection of technologies that will shape the nation's infrastructure in the future. Some of these technologies include advanced traffic monitoring, signal control systems, traveler information system, vehicle guidance and control systems, and commercial vehicle technologies like weigh in motion detector etc.

K-FACTOR – This factor considers the social or economic linkages between zones. Particularly considers those factors that are not considered by trip distribution model. K-factor is applied to the Gravity Model as a correction factor.

LANE GROUP – Lane group refers to the lane or lanes available for each movement at an intersection approach.

LEVEL OF SERVICE – A qualitative measure that describes the operational condition within a traffic stream, and their perception by motorists and/or passengers. A level of service definition generally describes these conditions in terms of such factors as speeds and travel time, freedom to maneuver, traffic interruptions, comfort and convenience and safety.

LINKS – Roadway or rail segments. Link information can include beginning and ending nodes, length , capacity, speed, area type, etc.

Glossary of Terms

LOSS TIME – Time during which the intersection is not effectively used by traffic from any direction. These times occur during a portion of the change interval (when the intersection is cleared), and at the beginning of each phase as the first few cars in a standing queue experience startup delay.

MAJOR INVESTMENT STUDY (MIS) – A study conducted to determine the cost/benefit of various alternative routes under consideration.

METROPOLITAN PLANNING AREA (MPA) – The geographic areas where the metropolitan transportation planning process must be carried out as required by Federal and State Laws.

METROPOLITAN PLANNING ORGANIZATION (MPO) – A metropolitan planning area's forum for cooperative transportation decision making.

MODE CHOICE – Predicts how the trips will be divided among the available modes of travel (ex. Car, Transit, Car Pool, etc.).

NETWORKS – Network representation include both highway and transit systems. A network is usually composed of nodes, links, centroids and external stations required for the study,

NODES – the endpoints of links, generally located at roadway intersections.

PEAK HOUR – The hour during which the maximum traffic occurs. The peak hour will vary from facility to facility and place to place.

PEAK-HOUR FACTOR – It is the ratio of total hourly volume to the maximum fifteen minute rate of flow within the hour.

PERMISSIVE TURNING MOVEMENT – Movements made through conflicting pedestrian traffic or opposing flow.

PHASE – The part of a cycle allocated to any combination of traffic movements receiving the right-of-way simultaneously during one or more intervals.

PRETIMED OPERATION – In a pre-timed operation, the cycle length, phases, green times and change intervals are all preset.

RATE OF FLOW – The equivalent hourly rate at which vehicles pass over a given point on a roadway during a given time interval less than one hour (usually 15 minutes).

Glossary of Terms

ROADWAY CONDITION – Refer to the geometric characteristics of the street. The types of facility and the development environment, the number of lanes (by direction), lane and shoulder widths, lateral clearances, and design speed.

SAMPLE – A sample is a collection of units selected from the population of interest to predict certain attributes of the population.

SEMI-ACTUATED OPERATION – In semi-actuated operation, the designated main street has a "green" phase until demand is on the side street or until a preset maximum side street green phase is reached.

SIGNALIZATION – Signalization includes a full definition of the signal phasing, timing, type of control, and an evaluation of signal progression on each approach.

SKIMS – Skims or travel time matrices contain information about the minimum impedance paths between all zone pairs in a network. These matrices are used in trip distribution, mode choice and traffic assignment steps of transportation forecasting modeling.

SPECIAL GENERATOR – A land use with unusually high or low generation characteristics. Example, Hospital, Schools, Military Bases and University.

SPEED – A rate of motion expressed as distance per unit of time, generally as miles per (mph) or kilometers per hour (km/h).

SPLIT TIMES – The amount of cycle time allocated for a particular phase.

TRANSIT CODING – The tracing of pat, producing and checking comparisons of transit and auto travel time, producing tables with number of transfers, wait-time, etc. to check reasonable values.

TRAFFIC ANALYSIS ZONES (TAZ's) – level of geographic detail used in most transportation planning application, particularly for modeling purposes. For a TAZ in modeling summarizes the socioeconomic characteristics of travel.

TREES – are a visual representation of the shortest path found in a network. The shortest path from a given zone to all other zones.

TRAFFIC CONDITIONS - Refer to the characteristics of the traffic stream using the facility. This is defined by the distribution of vehicle types in the traffic stream, and the directional distribution of traffic.

Glossary of Terms

TRAVEL DAIRY SURVEYS – Surveys that are applied to each member of the household that travels during the designated study period. Travel dairies are carried and completed by commuters during each day until the end of the study period.

TRIP ASSIGNMENT – Prediction of route choice, or the number of trips using highway links and transit links or lines.

TRIP DISTRIBUTION – Determines the destination choice of the trips and percentage of trips reaching a destination.

TRANSPORTATION DEMAND MANAGEMENT (TDM) - TDM program are primarily aimed at reducing peak period congestion through the reduction in the number of auto trips using programs such as parking charges, congestion pricing, toll roads, transit subsidies, tele-commuting and variable work hours.

TRIP GENERATION – Forecasting the number of trips that will be made on decisions as to the choice of trip frequency and landuse.

TRANSIT INVENTORY – The transit network requires data related to transit routes, headway's, location of transit stops, park and ride facilities, operational features and other transit related data for modeling purposes.

TRANSPORTATION IMPROVEMENT PROGRAM (TIP) – A staged, multi-year, inter-modal program of transportation projects developed by MPO's in cooperation with the state and transit operators.

TRIP LENGTH FREQUENCY DIAGRAM – The diagram provides the total number of trips and their trip lengths. This diagram is developed to compare observed data (home survey) with model output.

URBAN TRAVEL DEMAND FORECASTING – Refers to predicting travel behavior and resulting demand for a specific future time frame, based on several assumptions dealing with landuse, the number and character of trip makers, and the nature of transportation system.

VOLUME – The total number of vehicles that pass over a given point or section of a lane during a given time interval. Volume is expressed in terms of annual, daily and hourly.

ZONES – A geographic aggregations of individual households and businesses in the study area and this includes the study area.

Source: Highway Capacity Manual, Special Report 209, 1994.

Acronyms

AADT	Annual Average Daily Traffic
AASHTO	Association of State Highway and Transportation Officials
ADT	Average Daily Traffic
APTA	American Public Transit Association
AQMP	Air Quality Management Program
ASCE	American Society of Civil Engineers
ATES	Advanced Traffic Engineering Solution
ATIS	Advanced Traveler Information System
ATMS	Advanced Traffic Management System
AVL	Advanced Vehicle Location
AVR	Average Vehicle Ridership
BTES	Basic Traffic Engineering Solution
CAA	Clean Air Act
CAAA	Clean Air Act Amendments
CCTV	Closed Circuit Television
CO	Carbon Monoxide
EEMM	Environmental Enhancement and Mitigation Demand
ETC	Electronic Toll Collection
FTIP	Federal Transportation Improvement Program
GPS	Global Positioning System
HAR	Highway Advisory Radio
HBO	Home-Based Other
HBW	Home-Based Work (Trips)
HC	Hydrocarbons
HCM	Highway Capacity Manual
HOV	High Capacity Vehicle Lanes

Acronyms

ISTEA	Intermodal Surface Transportation Efficiency Act
ITE	Institute of Transportation Engineers
ITS	Intelligent Transportation System
LOS	Level of Service
MPO	Metropolitan Planning Organization
MTA	Metropolitan Transportation Authority
MUTCD	Manual of Uniform Traffic Control Devices
NHB	Non-Home Based
NIS	Network Information System
NOX	Nitrogen Oxides
PHF	Peak Hour Factor
ROC	Reactive Organic Compounds
RTMC	Regional Transportation Management Center
SIP	State Improvement Program
SITE	Site Impact Traffic Evaluation
SOV	Single Occupancy Vehicle Lane
STF	State Transportation Fund
TAZ	Traffic Analysis Zones
TCDI	Traffic Control Devices Inventory
TCM	Transportation Control Measures
TDF	Travel Demand Forecasting
TDM	Travel Demand Management
TEA	Transportation Equity Act
TIS	Traffic Impact Study
TLFD	Trip Length Frequency Diagram
TMC	Traffic Management Center
TMC	Turning Movement Counts
TOC	Traffic Operation Center

Acronyms

TSM	Traffic System Management System
TTF	Transportation Tax Fund
US	United States
WWW	World Wide Web

Abbreviations

SEC	Seconds
(g/c)	Green ratio for Lane Group
Aj	Attraction at Zone j
Ci	Capcity of Lane Group
COST	Out of Pocket Cost
d	Distance (feet)
E/W	East-West
EB	Eastbound
f	Coefficient of Friction
Fij	Friction Factor
g	Grade (%)
gr.	grams
H	Headway
IVT	In-Vehicle Time
Kij	Zone to Zone Attraction Factor
LT	Left Turn
mg.	Milligram
min.	Minutes
mph	Miles per Hour
N/S	North-South
NB	Northbound
OT	Other
OUT	Out of Vehicle Time
Pi	Production at Zone I
ppm	Parts per million
S	Shared Lane or Signalized
SB	Southbound

Abbreviations

Si	Saturation Flow Rate
TH	Through
Tij	Trips from Zone I to Zone j
Up	Transit Utility
Us	Auto Utility
US	Unsignalized
V	Initial Speed (mph)
vol.	Volume
vphg	Vehicles Per Hour of Green
WAIT	Wait Time
WB	Westbound
X	V/C Ratio for Lane Group
XFER	Transfer

Index

Index

High occupancy vehicles (HOV), 63,87
Highway Capacity Manual (HCM), 41, 60
Home based other, 67
Home based work, 67

Institute of Transportation Engineers (ITE), 95,99
Intelligent Transportation System (ITS), 117
Intersection delays, 26
ISTEA, 118
ITS architecture, 118

Level of service (LOS), 17, 41
Local funding, 115

Maintenance signs, 2
Manual on Uniform Traffic Control Devices (MUTCD), 1
Mode Choice, 67

National Ambient Air Quality Standards (NAAQS), 104
Non-home based, 67

Off-peak period, 6
Overall speed, 12

Parking area, 53
Parking control, 54
Parking demand, 53
Parking security, 54
Parking study, 52
Parking turnover, 56
Peak flow rate, 20
Peak hour factor (PHF), 20
Peak period, 6
Pedestrian study, 59
Percentage of vehicles stopped, 27
Poisson distribution, 44

Queuing analysis, 47

Recreational signs, 2
Regional transportation improvement program, 112
Regulatory signs, 2
Running speed, 12

Index